MATH

Grade 4

Thomas J. Richards
Mathematics Teacher
Lamar Junior-Senior High School
Lamar, Missouri

This book is dedicated to our children — Alyx, Nathan, Fred S., Dawn, Molly, Ellen, Rashaun, Brianna, Michele, Bradley, BriAnne, Kristie, Caroline, Dominic, Corey, Lindsey, Spencer, Morgan, Brooke, Cody, Sydney — and to all children who deserve a good education and who love to learn.

McGraw-Hill Consumer Products

McGraw-Hill
Consumer Products

A Division of The **McGraw·Hill** Companies

Copyright © 1998 McGraw-Hill Consumer Products.
Published by McGraw-Hill Learning Materials, an imprint of
McGraw-Hill Consumer Products.

Send all inquiries to:
McGraw Hill Consumer Products
8787 Orion Place
Columbus, OH 43240-4027

ISBN 1-57768-114-2

8 9 10 POH 03 02 01 00

Table of Contents

The SPECTRUM
Contents

MATHEMATICS Series
of Units

Using This Book

SPECTRUM MATHEMATICS is a non-graded, consumable series for students who need special help with the basic skills of computation and problem solving. This successful series emphasizes skill development and practice, without complex terminology or abstract symbolism. Because of the nature of the content and the students for whom the series is intended, readability has been carefully controlled to comply with the mathematics level of each book.

Features:

- A **Pre-Test** at the beginning of each chapter helps determine a student's understanding of the chapter content. The Pre-Test enables students and teachers to identify specific skills that need attention.

- **Developmental exercises** are provided at the top of the page when new skills are introduced. These exercises involve students in learning and serve as an aid for individualized instruction or independent study.

- **Abundant opportunities for practice** follow the developmental exercises.

- **Problem-solving pages** enable students to apply skills to realistic problems they will meet in everyday life.

- A **Test** at the end of each chapter gives students and teachers an opportunity to check understanding. A **Mid-Book Test**, covering Chapters 1–7, and a **Final Test**, covering all chapters, provide for further checks of understanding.

- A **Record of Test Scores** is provided on page xvi of this book so students can chart their progress as they complete each chapter test.

- **Answers** to all problems and test items are included at the back of the book.

This is the third edition of *SPECTRUM MATHEMATICS*. The basic books have remained the same. Some new, useful features have been added.

New Features:

- **Scope and Sequence Charts** for the entire Spectrum Mathematics series are included on pages iv–v.

- **Basic Facts Tests** for addition, subtraction, multiplication, and division are included on pages vii–xiv. There are two forms of each test. These may be given at any time the student or teacher decides they are appropriate.

- An **Assignment Record Sheet** is provided on page xv.

Addition Facts (Form A)

NAME _____

	a	b	c	d	e	f	g	h
1.	5 +7	0 +1	7 +2	4 +1	1 +6	3 +0	9 +5	8 +4
2.	5 +4	1 +5	8 +5	2 +9	4 +6	2 +0	6 +9	6 +3
3.	9 +4	1 +1	4 +2	6 +4	3 +9	9 +6	5 +3	0 +4
4.	2 +2	5 +6	3 +3	0 +0	7 +1	8 +3	6 +2	8 +0
5.	7 +3	0 +6	5 +2	7 +9	8 +2	9 +3	7 +7	2 +8
6.	6 +7	8 +6	4 +4	3 +4	2 +3	7 +0	5 +0	1 +3
7.	6 +5	9 +9	8 +7	5 +8	0 +9	7 +8	4 +9	3 +5
8.	5 +9	2 +5	7 +4	3 +6	4 +5	3 +8	7 +6	4 +7
9.	1 +2	5 +5	8 +8	9 +2	4 +8	9 +1	8 +9	2 +7
10.	3 +7	6 +8	2 +6	9 +8	6 +6	1 +8	9 +7	7 +5

Perfect score: 80   My score: _____

Addition Facts (Form B)

	a	*b*	*c*	*d*	*e*	*f*	*g*	*h*
1.	2 +1	8 +5	6 +5	4 +6	1 +0	7 +2	5 +5	3 +3
2.	8 +4	0 +0	7 +3	3 +5	9 +4	5 +1	2 +9	6 +2
3.	5 +6	2 +2	4 +7	0 +2	9 +0	8 +3	8 +7	3 +6
4.	3 +1	8 +2	5 +2	3 +2	6 +1	1 +1	6 +6	9 +9
5.	9 +3	1 +4	7 +4	5 +7	4 +8	0 +3	8 +1	2 +8
6.	0 +8	2 +4	6 +7	5 +3	9 +8	7 +9	3 +7	6 +0
7.	7 +5	9 +6	1 +7	2 +5	8 +6	5 +8	4 +9	0 +5
8.	4 +4	6 +8	6 +3	3 +8	8 +9	7 +8	2 +7	4 +5
9.	3 +9	9 +7	0 +7	5 +9	1 +9	6 +9	4 +3	7 +7
10.	5 +4	4 +0	8 +8	7 +6	9 +2	2 +6	9 +5	6 +4

Perfect score: 80   My score: _____

Subtraction Facts (Form A)

	a	b	c	d	e	f	g	h
1.	6 −2	1 1 −7	3 −3	1 3 −5	6 −1	1 1 −6	0 −0	1 3 −8
2.	6 −3	1 2 −6	5 −4	1 0 −6	7 −2	1 3 −9	4 −1	1 0 −7
3.	5 −2	1 0 −9	8 −2	1 2 −9	7 −3	1 1 −5	9 −6	1 7 −8
4.	8 −4	1 2 −5	9 −1	1 0 −3	9 −7	1 3 −4	1 0 −1	1 5 −9
5.	7 −5	1 1 −9	3 −1	1 4 −8	8 −3	1 0 −5	1 2 −3	1 0 −8
6.	2 −1	1 4 −5	6 −5	1 2 −8	7 −6	1 1 −3	1 0 −4	1 4 −7
7.	9 −9	1 2 −7	4 −0	1 3 −7	1 −1	1 6 −9	5 −5	1 5 −6
8.	8 −1	1 5 −8	9 −4	1 3 −6	7 −0	1 1 −2	9 −3	1 6 −7
9.	9 −2	1 4 −6	6 −0	1 7 −9	8 −8	1 2 −4	1 0 −2	1 1 −8
10.	9 −0	1 8 −9	4 −2	1 5 −7	2 −0	1 6 −8	1 1 −4	1 4 −9

Perfect score: 80   My score: _____

Subtraction Facts (Form B)

	a	b	c	d	e	f	g	h
1.	1 2 −3	9 −5	1 5 −6	7 −3	1 0 −4	9 −2	5 −5	1 2 −6
2.	1 1 −2	8 −4	1 0 −5	7 −7	1 6 −8	8 −0	1 0 −3	1 0 −9
3.	1 7 −8	6 −3	1 6 −9	5 −3	1 0 −7	8 −2	9 −4	1 4 −5
4.	1 1 −3	9 −8	1 1 −7	6 −6	1 3 −9	2 −1	7 −2	1 3 −6
5.	1 5 −9	2 −2	1 1 −5	0 −0	1 0 −8	6 −5	1 3 −4	1 2 −7
6.	1 0 −2	7 −4	1 2 −9	4 −4	1 0 −1	8 −7	3 −0	1 5 −8
7.	1 1 −4	1 −0	1 4 −8	4 −3	1 7 −9	9 −3	1 3 −7	1 1 −8
8.	1 4 −7	8 −6	1 1 −6	5 −0	1 2 −5	6 −4	5 −1	1 6 −7
9.	1 1 −9	7 −1	1 3 −8	3 −2	1 0 −6	9 −6	1 8 −9	1 3 −5
10.	1 2 −4	9 −9	1 4 −6	8 −5	1 5 −7	4 −2	1 2 −8	1 4 −9

Perfect score: 80   My score: _____

x

Multiplication Facts (Form A)

	a	*b*	*c*	*d*	*e*	*f*	*g*	*h*
1.	8 ×2	2 ×3	7 ×3	4 ×5	3 ×1	5 ×2	2 ×9	7 ×0
2.	5 ×1	8 ×0	1 ×1	4 ×6	2 ×0	9 ×1	6 ×3	0 ×9
3.	4 ×7	0 ×0	7 ×1	3 ×3	8 ×3	5 ×3	4 ×0	9 ×2
4.	2 ×4	3 ×9	6 ×2	9 ×7	0 ×1	8 ×4	1 ×7	2 ×2
5.	9 ×8	1 ×4	8 ×5	5 ×4	7 ×9	6 ×1	7 ×4	2 ×8
6.	5 ×5	8 ×6	4 ×8	7 ×5	6 ×4	3 ×5	6 ×9	9 ×6
7.	1 ×9	7 ×8	0 ×5	8 ×7	1 ×2	5 ×8	3 ×8	6 ×0
8.	5 ×9	2 ×7	8 ×8	9 ×9	9 ×5	6 ×5	8 ×9	5 ×7
9.	9 ×4	7 ×6	3 ×6	6 ×8	1 ×8	0 ×6	4 ×3	9 ×3
10.	3 ×7	0 ×7	4 ×4	5 ×6	4 ×9	2 ×6	7 ×7	6 ×6

Perfect score: 80   **My score:** _____

Multiplication Facts (Form B)

	a	*b*	*c*	*d*	*e*	*f*	*g*	*h*
1.	3 ×8	2 ×2	7 ×3	3 ×2	0 ×6	6 ×3	2 ×7	4 ×1
2.	3 ×3	0 ×0	6 ×2	2 ×1	5 ×2	3 ×0	7 ×4	1 ×0
3.	5 ×4	7 ×2	5 ×0	4 ×2	3 ×7	8 ×1	0 ×4	5 ×5
4.	3 ×9	2 ×8	1 ×1	5 ×3	7 ×5	4 ×9	8 ×9	1 ×9
5.	9 ×4	5 ×6	8 ×5	4 ×8	0 ×3	8 ×6	6 ×4	9 ×9
6.	1 ×3	9 ×2	2 ×9	7 ×6	9 ×8	5 ×7	4 ×3	0 ×8
7.	7 ×9	4 ×7	8 ×4	5 ×8	4 ×4	7 ×1	9 ×0	3 ×6
8.	6 ×8	3 ×4	0 ×2	9 ×3	1 ×5	7 ×7	6 ×5	8 ×2
9.	8 ×8	1 ×6	4 ×5	6 ×6	5 ×9	9 ×5	6 ×9	7 ×0
10.	4 ×6	6 ×7	9 ×7	7 ×8	8 ×3	3 ×5	1 ×8	9 ×6

Perfect score: 80   My score: _____

Division Facts (Form A)

	a	b	c	d	e	f	g
1.	5$\overline{)10}$	7$\overline{)7}$	6$\overline{)30}$	8$\overline{)24}$	4$\overline{)12}$	9$\overline{)9}$	5$\overline{)40}$
2.	9$\overline{)45}$	4$\overline{)8}$	5$\overline{)15}$	7$\overline{)21}$	6$\overline{)12}$	2$\overline{)18}$	8$\overline{)32}$
3.	2$\overline{)4}$	8$\overline{)16}$	4$\overline{)28}$	9$\overline{)54}$	3$\overline{)18}$	7$\overline{)56}$	3$\overline{)0}$
4.	5$\overline{)5}$	1$\overline{)3}$	9$\overline{)0}$	7$\overline{)63}$	5$\overline{)20}$	6$\overline{)36}$	4$\overline{)32}$
5.	9$\overline{)36}$	4$\overline{)24}$	3$\overline{)6}$	8$\overline{)40}$	2$\overline{)6}$	3$\overline{)12}$	9$\overline{)63}$
6.	8$\overline{)8}$	1$\overline{)5}$	6$\overline{)42}$	1$\overline{)6}$	3$\overline{)15}$	8$\overline{)72}$	1$\overline{)7}$
7.	1$\overline{)2}$	6$\overline{)18}$	2$\overline{)16}$	4$\overline{)36}$	2$\overline{)8}$	1$\overline{)9}$	7$\overline{)0}$
8.	9$\overline{)18}$	1$\overline{)1}$	7$\overline{)49}$	8$\overline{)48}$	4$\overline{)20}$	9$\overline{)27}$	5$\overline{)25}$
9.	2$\overline{)10}$	6$\overline{)48}$	3$\overline{)9}$	9$\overline{)72}$	2$\overline{)0}$	7$\overline{)28}$	3$\overline{)3}$
10.	5$\overline{)35}$	7$\overline{)42}$	2$\overline{)14}$	4$\overline{)0}$	6$\overline{)54}$	7$\overline{)14}$	3$\overline{)24}$
11.	1$\overline{)8}$	6$\overline{)0}$	3$\overline{)27}$	5$\overline{)45}$	3$\overline{)21}$	2$\overline{)2}$	8$\overline{)56}$
12.	2$\overline{)12}$	9$\overline{)81}$	8$\overline{)64}$	5$\overline{)30}$	7$\overline{)35}$	4$\overline{)4}$	6$\overline{)24}$

Perfect score: 84   My score: _____

Division Facts (Form B)

NAME _____

	a	b	c	d	e	f	g
1.	$1\overline{)8}$	$9\overline{)9}$	$3\overline{)15}$	$7\overline{)63}$	$2\overline{)18}$	$5\overline{)20}$	$8\overline{)16}$
2.	$4\overline{)28}$	$5\overline{)25}$	$6\overline{)6}$	$2\overline{)0}$	$9\overline{)18}$	$3\overline{)18}$	$6\overline{)48}$
3.	$2\overline{)16}$	$6\overline{)54}$	$7\overline{)56}$	$5\overline{)15}$	$5\overline{)45}$	$1\overline{)9}$	$8\overline{)8}$
4.	$6\overline{)0}$	$2\overline{)2}$	$6\overline{)18}$	$3\overline{)0}$	$9\overline{)27}$	$4\overline{)0}$	$4\overline{)36}$
5.	$1\overline{)7}$	$3\overline{)21}$	$8\overline{)24}$	$7\overline{)14}$	$2\overline{)4}$	$8\overline{)48}$	$1\overline{)2}$
6.	$8\overline{)56}$	$7\overline{)49}$	$4\overline{)32}$	$3\overline{)12}$	$6\overline{)24}$	$2\overline{)6}$	$9\overline{)36}$
7.	$3\overline{)3}$	$9\overline{)81}$	$1\overline{)3}$	$8\overline{)0}$	$5\overline{)10}$	$1\overline{)0}$	$7\overline{)21}$
8.	$7\overline{)0}$	$5\overline{)30}$	$4\overline{)24}$	$2\overline{)8}$	$3\overline{)24}$	$8\overline{)40}$	$4\overline{)16}$
9.	$2\overline{)10}$	$7\overline{)42}$	$9\overline{)72}$	$6\overline{)30}$	$7\overline{)28}$	$6\overline{)42}$	$8\overline{)72}$
10.	$4\overline{)20}$	$1\overline{)5}$	$3\overline{)27}$	$3\overline{)9}$	$5\overline{)35}$	$2\overline{)12}$	$9\overline{)45}$
11.	$2\overline{)14}$	$7\overline{)7}$	$8\overline{)32}$	$9\overline{)63}$	$6\overline{)12}$	$5\overline{)0}$	$4\overline{)12}$
12.	$1\overline{)4}$	$4\overline{)4}$	$8\overline{)64}$	$7\overline{)35}$	$6\overline{)36}$	$9\overline{)54}$	$5\overline{)40}$

Perfect score: 84   My score: _____

Assignment Record Sheet

NAME _____

Pages Assigned	Date	Score

Pages Assigned	Date	Score

Pages Assigned	Date	Score

SPECTRUM MATHEMATICS

Record of Test Scores

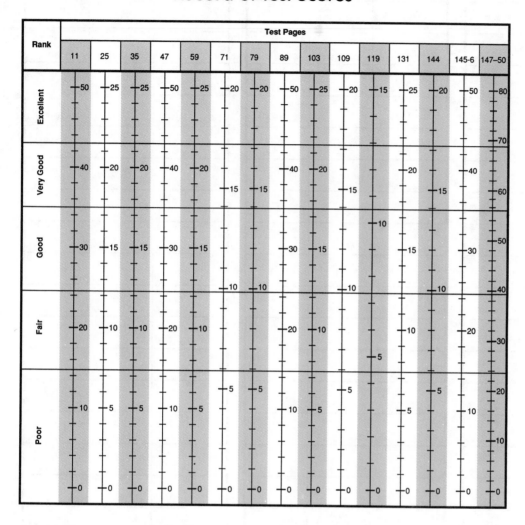

To record the score you receive on a TEST:

(1) Find the vertical scale below the page number of that TEST,
(2) on that vertical scale, draw a ● at the mark which represents your score.

For example, if your score for the TEST on page 11 is "My score: 32," draw a ● at the 32-mark on the first vertical scale. A score of 32 would show that your rank is "Good." You can check your progress from one test to the next by connecting the dots with a line segment.

PRE-TEST—Addition and Subtraction

Add.

	a	*b*	*c*	*d*	*e*	*f*
1.	45 +4	73 +5	64 +5	22 +7	42 +5	74 +4
2.	4 +43	3 +54	6 +43	3 +34	5 +24	2 +46
3.	70 +10	60 +20	40 +50	40 +30	70 +20	30 +40
4.	54 +21	23 +41	11 +16	22 +23	45 +21	53 +32
5.	68 +30	11 +25	34 +30	12 +80	12 +71	45 +40

Subtract.

	a	*b*	*c*	*d*	*e*	*f*
6.	28 −7	57 −6	46 −4	29 −8	47 −5	78 −6
7.	50 −10	40 −20	60 −40	70 −30	90 −60	80 −30
8.	89 −46	97 −94	89 −55	78 −23	98 −48	67 −23
9.	78 −70	98 −32	99 −21	67 −61	89 −10	66 −51

Perfect score: 54   My score: _____

1

Problem-Solving Pre-Test

Solve each problem.

1. Heather took 3 pictures of sailboats and 2 pictures of fishing boats. How many pictures of boats did she take in all?

She took _____ pictures of sailboats.

She took _____ pictures of fishing boats.

She took _____ pictures of boats in all.

1.

2. Heather also took 9 pictures of people swimming. Only 5 of the pictures had sea gulls in them. How many pictures did not have sea gulls in them?

There were _____ pictures of people swimming.

_____ of the pictures had sea gulls in them.

_____ of the pictures did not have sea gulls in them.

2.

3. During the day Heather used 4 rolls of black-and-white film and 3 rolls of color film. How many rolls of film did she use in all?

She used _____ rolls of black-and-white film.

She used _____ rolls of color film.

She used _____ rolls of film in all.

3.

Perfect score: 9   My score: _____

2

Lesson 1 Addition

4 → Find the **4**-row.

+5 → Find the **5**-column.

9 ← The sum is named where the 4-row and 5-column meet.

Use the table to add.

7
+8

5-column →

+	0	1	2	3	4	5	6	7	8	9
0	0	1	2	3	4	5	6	7	8	9
1	1	2	3	4	5	6	7	8	9	10
2	2	3	4	5	6	7	8	9	10	11
3	3	4	5	6	7	8	9	10	11	12
4	4	5	6	7	8	9	10	11	12	13
5	5	6	7	8	9	10	11	12	13	14
6	6	7	8	9	10	11	12	13	14	15
7	7	8	9	10	11	12	13	14	15	16
8	8	9	10	11	12	13	14	15	16	17
9	9	10	11	12	13	14	15	16	17	18

4-row →

Add.

	a	b	c	d	e	f	g	h
1.	5 +3	2 +5	5 +4	6 +3	3 +4	3 +5	2 +7	3 +7
2.	6 +2	3 +6	3 +2	2 +4	4 +3	2 +6	7 +2	4 +4
3.	2 +3	0 +7	3 +1	9 +0	1 +8	0 +5	4 +2	6 +1
4.	7 +9	6 +7	9 +4	8 +3	4 +9	7 +3	8 +5	8 +9
5.	7 +7	9 +5	4 +7	6 +5	2 +8	7 +5	4 +8	7 +6
6.	9 +6	5 +8	5 +9	6 +6	9 +8	7 +4	3 +9	2 +9
7.	4 +6	1 +9	5 +7	9 +3	3 +8	8 +4	9 +7	9 +9

Perfect score: 56   My score: _____

3

Lesson 2   Subtraction

NAME _____

9 { Find the 9 in

−6 { the **6**-column.

3 ← The difference is named in the ▨ at the end of this row.

Use the table to subtract.

12
−4

−	0	1	2	3	4	5	6	7	8	9
0	0	1	2	3	4	5	6	7	8	9
1	1	2	3	4	5	6	7	8	9	10
2	2	3	4	5	6	7	8	9	10	11
3	3	4	5	6	7	8	9	10	11	12
4	4	5	6	7	8	9	10	11	12	13
5	5	6	7	8	9	10	11	12	13	14
6	6	7	8	9	10	11	12	13	14	15
7	7	8	9	10	11	12	13	14	15	16
⑧	8	9	10	11	12	13	14	15	16	17
9	9	10	11	12	13	14	15	16	17	18

Subtract.

	a	b	c	d	e	f	g	h
1.	7 −6	9 −7	7 −4	3 −2	8 −0	9 −5	8 −7	7 −1
2.	5 −3	6 −5	7 −0	8 −2	8 −4	9 −9	8 −5	5 −2
3.	13 −6	10 −2	14 −5	12 −5	11 −6	10 −9	18 −9	11 −9
4.	15 −9	11 −8	16 −8	12 −6	14 −6	11 −3	15 −8	12 −8
5.	16 −9	13 −4	15 −7	10 −4	12 −9	13 −5	11 −7	10 −6
6.	14 −9	11 −2	10 −3	14 −8	16 −7	14 −7	17 −8	11 −4
7.	13 −9	13 −7	10 −8	15 −6	13 −8	12 −7	17 −9	11 −5

Perfect score: 56   My score: _____

4

Lesson 3 Addition and Subtraction

	Add the ones.	Add the tens.		Subtract the ones.	Subtract the tens.
36 +2	3**6** +2 ___ 8	36 +2 ___ 38	57 −4 ___	57 −4 ___ 3	57 −4 ___ 53

Add.

	a	b	c	d	e	f
1.	20 +7	31 +1	42 +1	18 +1	21 +6	54 +4
2.	22 +1	42 +2	63 +3	75 +4	64 +5	96 +3
3.	3 +66	2 +16	2 +17	1 +28	9 +30	7 +32
4.	7 +41	8 +61	2 +35	6 +41	8 +40	5 +71

Subtract.

5.	19 −8	38 −7	59 −5	27 −6	45 −3	96 −5
6.	97 −5	49 −6	18 −5	69 −7	38 −6	59 −9
7.	89 −4	28 −4	46 −4	77 −4	89 −3	65 −2
8.	79 −1	68 −2	39 −2	86 −3	57 −3	78 −3

Perfect score: 48   My score: _____

Problem Solving

Solve each problem.

1. Regina worked 27 hours. Louis worked 6. How many more hours did Regina work than Louis?

Regina worked _____ hours.

Louis worked _____ hours.

Regina worked _____ more hours.

2. The Cubs got 8 hits in the first game. They got 11 hits in the second game. How many hits did they get in both games?

They got _____ hits in the first game.

They got _____ hits in the second game.

They got _____ hits in both games.

3. Five of Mr. Spangler's workers are absent today. Twenty-three of them are at work today. How many workers does he have in all?

_____ workers are present.

_____ workers are absent.

He has _____ workers.

4. Melody has 12 records. Paul has 6 records. How many records do they have in all?

Melody has _____ records.

Paul has _____ records.

They have _____ records in all.

5. There are 19 women and 7 men in Mildred's apartment building. How many more women than men live in that apartment building?

_____ more women live in the building.

1.

2.

3.

4.   5.

Perfect score: 13   My score: _____

6

Lesson 4 Addition

```
   6          6 0
  +2         +2 0
  ---        -----
   8          8 0
```

If 6 + 2 = 8,
then 60 + 20 = 80.

	Add the ones.	Add the tens.
36	3 6	3 6
+21	+2 1	+2 1
	----	----
	7	5 7

Add.

	a	b	c	d	e	f
1.	5 +1	5 0 +1 0	4 +3	4 0 +3 0	7 +2	7 0 +2 0
2.	3 +6	3 0 +6 0	5 +2	5 0 +2 0	1 +8	1 0 +8 0
3.	6 0 +3 0	4 0 +4 0	7 0 +1 0	2 0 +6 0	5 0 +3 0	4 0 +2 0
4.	1 7 +4 1	2 8 +6 1	3 2 +3 5	4 6 +4 1	2 8 +4 0	1 5 +7 1
5.	2 2 +1 4	3 3 +6 4	4 4 +3 3	8 5 +1 3	7 1 +2 4	7 1 +1 3
6.	2 1 +7 3	2 4 +6 2	2 5 +7 2	1 6 +8 2	4 2 +4 3	5 3 +4 2
7.	6 3 +3 6	5 8 +3 1	6 4 +2 4	4 7 +4 2	7 6 +1 3	3 5 +6 3

Perfect score: 42   My score: _____

7

Problem Solving

Solve each problem.

1. Eileen's family used 14 liters of milk this week. Last week they used 15 liters. How many liters of milk did they use in the two weeks?

_____ liters were used this week.

_____ liters were used last week.

_____ liters were used in the two weeks.

2. Tom drove 42 kilometers today. He drove 46 kilometers yesterday. How many kilometers did he drive during the two days?

_____ kilometers were driven today.

_____ kilometers were driven yesterday.

_____ kilometers were driven both days.

3. Our family has two dogs. Pepper weighs 31 pounds. Salt weighs 28 pounds. What is the combined weight of the two dogs?

Pepper weighs _____ pounds.

Salt weighs _____ pounds.

The combined weight is _____ pounds.

4. Leslie scored 43 points. Michael scored 25 points. How many points did they score?

Leslie scored _____ points.

Michael scored _____ points.

They scored _____ points.

5. Mr. Cook was 25 years old when Mary was born. How old will he be when Mary has her thirteenth birthday?

Mr. Cook will be _____ years old.

1.

2.

3.

4.

5.

Perfect score: 13   My score: _____

8

Lesson 5   Subtraction

		Subtract the ones.	Subtract the tens.

$$\begin{array}{r} 5 \\ -3 \\ \hline 2 \end{array} \qquad \begin{array}{r} 50 \\ -30 \\ \hline 20 \end{array}$$

If $5 - 3 = 2$,
then $50 - 30 = 20$.

$$\begin{array}{r} 67 \\ -53 \\ \hline \end{array} \qquad \begin{array}{r} 67 \\ -53 \\ \hline 4 \end{array} \qquad \begin{array}{r} 67 \\ -53 \\ \hline 14 \end{array}$$

Subtract.

	a	b	c	d	e	f
1.	$\begin{array}{r} 4 \\ -2 \\ \hline \end{array}$	$\begin{array}{r} 40 \\ -20 \\ \hline \end{array}$	$\begin{array}{r} 7 \\ -4 \\ \hline \end{array}$	$\begin{array}{r} 70 \\ -40 \\ \hline \end{array}$	$\begin{array}{r} 9 \\ -6 \\ \hline \end{array}$	$\begin{array}{r} 90 \\ -60 \\ \hline \end{array}$
2.	$\begin{array}{r} 8 \\ -3 \\ \hline \end{array}$	$\begin{array}{r} 80 \\ -30 \\ \hline \end{array}$	$\begin{array}{r} 7 \\ -5 \\ \hline \end{array}$	$\begin{array}{r} 70 \\ -50 \\ \hline \end{array}$	$\begin{array}{r} 6 \\ -2 \\ \hline \end{array}$	$\begin{array}{r} 60 \\ -20 \\ \hline \end{array}$
3.	$\begin{array}{r} 40 \\ -30 \\ \hline \end{array}$	$\begin{array}{r} 90 \\ -70 \\ \hline \end{array}$	$\begin{array}{r} 80 \\ -40 \\ \hline \end{array}$	$\begin{array}{r} 70 \\ -20 \\ \hline \end{array}$	$\begin{array}{r} 60 \\ -40 \\ \hline \end{array}$	$\begin{array}{r} 50 \\ -10 \\ \hline \end{array}$
4.	$\begin{array}{r} 98 \\ -87 \\ \hline \end{array}$	$\begin{array}{r} 76 \\ -66 \\ \hline \end{array}$	$\begin{array}{r} 89 \\ -61 \\ \hline \end{array}$	$\begin{array}{r} 57 \\ -45 \\ \hline \end{array}$	$\begin{array}{r} 49 \\ -39 \\ \hline \end{array}$	$\begin{array}{r} 65 \\ -53 \\ \hline \end{array}$
5.	$\begin{array}{r} 69 \\ -40 \\ \hline \end{array}$	$\begin{array}{r} 78 \\ -45 \\ \hline \end{array}$	$\begin{array}{r} 69 \\ -37 \\ \hline \end{array}$	$\begin{array}{r} 45 \\ -22 \\ \hline \end{array}$	$\begin{array}{r} 87 \\ -83 \\ \hline \end{array}$	$\begin{array}{r} 95 \\ -60 \\ \hline \end{array}$
6.	$\begin{array}{r} 53 \\ -51 \\ \hline \end{array}$	$\begin{array}{r} 94 \\ -51 \\ \hline \end{array}$	$\begin{array}{r} 84 \\ -44 \\ \hline \end{array}$	$\begin{array}{r} 98 \\ -43 \\ \hline \end{array}$	$\begin{array}{r} 27 \\ -12 \\ \hline \end{array}$	$\begin{array}{r} 66 \\ -20 \\ \hline \end{array}$
7.	$\begin{array}{r} 86 \\ -21 \\ \hline \end{array}$	$\begin{array}{r} 74 \\ -10 \\ \hline \end{array}$	$\begin{array}{r} 95 \\ -31 \\ \hline \end{array}$	$\begin{array}{r} 83 \\ -12 \\ \hline \end{array}$	$\begin{array}{r} 49 \\ -22 \\ \hline \end{array}$	$\begin{array}{r} 78 \\ -50 \\ \hline \end{array}$

Perfect score: 42   My score: _____

Problem Solving

Solve each problem.

1. Catherine weighs 39 kilograms. Her sister weighs 23 kilograms. How much more does Catherine weigh?

Catherine weighs _____ kilograms.

Her sister weighs _____ kilograms.

Catherine weighs _____ kilograms more.

2. Bonnie can kick a football 28 yards. She can throw it 21 yards. How much farther can she kick the football?

She can kick the football _____ yards.

She can throw the football _____ yards.

She can kick the football _____ yards farther.

3. Marcos has 47 dollars. He plans to spend 25 dollars on presents. How much money will he have left?

Marcos has _____ dollars.

He plans to spend _____ dollars.

He will have _____ dollars left.

4. Jerry is running in a 26-mile race. So far he has run 16 miles. How many more miles must he run?

The race is _____ miles long.

Jerry has gone _____ miles.

He has _____ miles more to go.

5. Alene read 24 pages on Monday. She read 37 pages on Tuesday. How many more pages did she read on Tuesday than on Monday?

Alene read _____ more pages on Tuesday.

1.

2.

3.

4.   **5.**

Perfect score: 13   My score: _____

10

CHAPTER 1 TEST

Add.

	a	b	c	d	e	f
1.	20 +9	28 +1	51 +7	52 +7	42 +7	23 +6
2.	5 +31	2 +81	3 +31	3 +94	4 +31	7 +51
3.	40 +20	10 +60	20 +60	40 +20	30 +20	50 +30
4.	35 +60	45 +22	43 +40	45 +33	41 +55	35 +14
5.	62 +34	75 +13	73 +25	64 +23	43 +25	25 +74

Subtract.

	a	b	c	d	e
6.	68 −6	29 −1	47 −6	59 −2	78 −5
7.	70 −30	60 −50	90 −40	80 −30	60 −40
8.	88 −78	77 −34	98 −50	67 −41	98 −91
9.	96 −62	88 −23	76 −23	99 −78	87 −40

Perfect score: 50   My score: _____

PRE-TEST—Addition and Subtraction

Add.

	a	b	c	d	e	f
1.	1 2 +5	6 4 +2	7 8 +5	3 7 +5	8 7 +9	8 5 +4
2.	28 +7	42 +9	69 +5	7 +63	8 +47	7 +49
3.	67 +25	34 +72	87 +29	36 +215	295 +42	388 +25
4.	52 24 +37	62 57 +93	48 63 +75	96 52 +48	46 63 +72	96 51 +39

Subtract.

	a	b	c	d	e	f
5.	26 −7	83 −9	67 −8	43 −7	94 −6	76 −9
6.	43 −28	64 −25	73 −47	83 −24	76 −29	83 −25
7.	129 −67	839 −97	425 −84	987 −95	654 −57	832 −51
8.	364 −76	524 −69	608 −59	832 −47	784 −85	477 −98
9.	584 −75	834 −15	679 −89	983 −91	201 −96	485 −98

Perfect score: 54   My score: _____

12

Lesson 1 Addition

```
  3 ⟍
  4 ⟶ 7
 +5 — +5
      ——
       12
```

```
  3 ⟍
  1 ⟶ 4 ⟍
  2 — 2 ⟶ 6
 +5 — +5 — +5
              ——
               11
```

Add.

	a	b	c	d	e	f	g	h
1.	1 5 +7	1 8 +3	7 2 +5	4 5 +9	7 1 +9	2 2 +8	7 2 +8	6 3 +5
2.	3 2 +5	1 7 +5	6 2 +5	2 4 +5	3 3 +9	3 5 +7	4 4 +5	3 6 +4
3.	1 4 +8	5 3 +8	2 6 +8	4 1 +9	2 5 +4	5 4 +2	3 5 +3	6 3 +2
4.	2 3 3 +5	1 3 4 +8	6 2 1 +7	5 3 1 +4	4 4 1 +8	7 1 1 +9	6 2 1 +7	2 3 4 +9
5.	1 2 6 +9	1 3 5 +8	2 1 5 +7	2 2 5 +8	3 2 4 +7	3 1 2 +9	4 1 4 +6	2 3 2 +3
6.	3 5 2 +7	4 6 1 +9	3 5 5 +6	5 4 3 +7	7 1 2 +6	4 3 3 +8	8 3 1 +6	7 6 8 +5

Perfect score: 48   My score: _____

13

Problem Solving

Solve each problem.

1. A football team gained 2 yards on first down, 3 yards on second down, and 6 yards on third down. How many yards did they gain in the three downs?

They gained _____ yards on first down.

They gained _____ yards on second down.

They gained _____ yards on third down.

They gained _____ yards in the three downs.

2. Mr. Bernardo bought 5 apples, 4 oranges, and 6 pears. How many pieces of fruit did he buy?

He bought _____ apples.

He bought _____ oranges.

He bought _____ pears.

He bought _____ pieces of fruit.

3. During a hike, Trixie counted 3 maple trees, 4 oak trees, 2 poplar trees, and 7 aspen trees. How many trees did Trixie count?

Trixie counted _____ trees.

4. During a game, Amy made 4 baskets, Sally made 2, Tina made 3, and Kim made 6. How many baskets did these girls make?

The girls made _____ baskets.

5. One piece of rope is 3 meters long. Another is 6 meters long. A third piece is 2 meters long. What is the combined length of the three pieces?

The combined length is _____ meters.

6. Carlotta was absent from school 4 days in March, 6 days in April, 0 days in May, and 3 days in June. How many days was she absent during the four months?

Carlotta was absent _____ days.

| 1. |
| 2. |

| 3. | 4. |

| 5. | 6. |

Perfect score: 12   My score: _____

14

Lesson 2   Addition

Add the ones.   Add the tens.

```
  46              ¹            ¹
 +27             46          46
                +27         +27
 6 + 7 = 13 or 10 + 3   ───         ───
                  3          73
```

Add.

	a	b	c	d	e	f
1.	17 +9	14 +8	32 +9	25 +6	31 +9	53 +8
2.	3 +28	7 +64	9 +56	5 +37	4 +38	8 +65
3.	32 +28	78 +15	65 +16	74 +18	29 +55	65 +27
4.	25 +46	27 +14	16 +25	17 +33	17 +26	28 +44
5.	38 +45	47 +39	63 +28	39 +52	44 +29	73 +18
6.	37 +36	38 +19	45 +29	58 +19	16 +27	19 +27
7.	35 +48	16 +75	32 +49	27 +38	29 +58	51 +29

Perfect score: 42   My score: _____

15

Problem Solving

Solve each problem.

1. The station has 16 white-wall tires. It has 9 black-wall tires. How many tires does the station have?

There are _____ white-wall tires.

There are _____ black-wall tires.

There are _____ tires in all.

2. It takes 8 minutes to grease a car. It takes 15 minutes to change the oil. How long would it take to grease a car and change the oil?

It would take _____ minutes to grease the car and change the oil.

3. This morning 26 people bought gasoline. This afternoon 37 people bought gasoline. How many people have bought gasoline today?

_____ people have bought gasoline today.

4. Mrs. Verdugo bought 54 liters of gasoline on Tuesday. She bought 38 liters of gasoline on Friday. How many liters did she buy in all?

She bought _____ liters of gasoline.

1.

2.

3.   **4.**

Perfect score: 6   My score: _____

16

Lesson 3   Addition

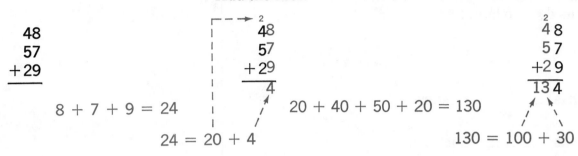

Add the ones.

Add the tens.

$$8 + 7 + 9 = 24$$

$$24 = 20 + 4$$

$$20 + 40 + 50 + 20 = 130$$

$$130 = 100 + 30$$

Add.

	a	b	c	d	e	f
1.	7 3 +5 2	6 4 +9 3	6 5 +6 4	5 1 +7 8	8 2 +3 4	6 0 +5 7
2.	4 7 +8 1	5 6 +8 2	7 8 +4 1	8 4 +9 2	7 6 +8 3	8 6 +5 3
3.	2 8 +7 3	3 9 +9 2	8 7 +7 3	9 9 +5 1	7 9 +5 3	5 6 +7 5
4.	6 2 5 4 +7 2	7 2 8 3 +5 1	9 5 6 1 +2 2	9 5 8 2 +7 1	5 2 2 1 +6 6	3 2 2 3 +9 4
5.	6 2 7 5 +4 3	8 3 7 5 +6 4	9 6 8 7 +4 2	8 6 7 8 +5 4	6 7 3 2 +4 5	8 4 2 3 +5 7
6.	6 1 2 2 +1 7	6 3 5 7 +8 3	9 6 4 8 +7 6	2 7 8 3 +9 1	8 1 8 9 +9 5	7 5 3 6 +2 4

Perfect score: 36   My score: _____

17

Problem Solving

Solve each problem.

1. Mr. Ford has 61 hens and 54 roosters. How many chickens does he have?

He has _____ hens.

He has _____ roosters.

He has _____ chickens.

2. Last week Tony read a 76-page book. This week he read an 83-page book. How many pages are in the two books?

Last week he read a _____ -page book.

This week he read an _____ -page book.

There are _____ pages in the two books.

3. Jackie has 75 feet of kite string. Mike has 96 feet of kite string. How much kite string do they have?

They have _____ feet of kite string.

4. A train went 57 kilometers the first hour. It went 65 kilometers the second hour. It went 52 kilometers the third hour. How far did it go?

The train went _____ kilometers.

5. A restaurant used 76 kilograms of potatoes, 14 kilograms of carrots, and 68 kilograms of meat. What was the weight of the three items?

The weight was _____ kilograms.

6. A cafeteria served 52 women, 47 men, and 69 children. How many people did the cafeteria serve?

_____ people were served.

1.
2.

3.	4.

5.	6.

Perfect score: 10   My score: _____

18

Lesson 4 Subtraction

To subtract the ones,
rename 4 tens and 7 ones
as "3 tens and 17 ones."

$$147 \quad \overset{3\ 17}{14\!\!\!/7}$$
$$-19 \qquad -19$$

Subtract
the ones.

$$\overset{3\ 17}{14\!\!\!/7}$$
$$-19$$
$$\overline{8}$$

Subtract
the tens.

$$\overset{3\ 17}{14\!\!\!/7}$$
$$-19$$
$$\overline{28}$$

Subtract
the hundreds.

$$\overset{3\ 17}{14\!\!\!/7}$$
$$-19$$
$$\overline{128}$$

Subtract.

	a	b	c	d	e	f
1.	26 −8	63 −7	72 −9	42 −5	83 −6	75 −7
2.	23 −17	65 −16	83 −27	32 −17	64 −37	70 −26
3.	64 −37	84 −25	42 −24	78 −39	47 −38	63 −47
4.	153 −27	625 −16	171 −25	835 −19	135 −16	635 −26
5.	543 −26	175 −28	483 −75	125 −19	785 −49	135 −27
6.	524 −15	684 −25	895 −26	768 −29	726 −17	325 −16
7.	385 −76	735 −17	633 −15	354 −45	576 −47	880 −35

Perfect score: 42   My score: _____

Problem Solving

Solve each problem.

1. A candle was 32 centimeters long when lit. When blown out, it was 19 centimeters long. How long was the part that burned away?

The candle was _____ centimeters long when lit.

After burning, it was _____ centimeters long.

_____ centimeters of the candle burned away.

2. A gasoline tank can hold 92 liters. It took 68 liters to fill the tank. How many liters were in the tank before filling it?

The tank can hold _____ liters.

It took _____ liters to fill the tank.

_____ liters were in the tank before filling.

3. Joy exercised 42 minutes this morning. She exercised 27 minutes this afternoon. How many more minutes did she exercise this morning?

Joy exercised _____ minutes this morning.

She exercised _____ minutes this afternoon.

Joy exercised _____ minutes more this morning.

4. There are 283 pupils at Adams School. Sixty-seven of these pupils were in the park program last year. How many were not in the program?

_____ pupils were not in the program.

5. Pablo's father weighs 173 pounds. Pablo weighs 65 pounds. How much more does Pablo's father weigh?

Pablo's father weighs _____ pounds more.

1.

2.

3.

4.

5.

Perfect score: 11   My score: _____

20

Lesson 5   Subtraction

Rename 306 as "2 hundreds, 10 tens, and 6 ones."		Then rename as "2 hundreds, 9 tens, and 16 ones." Subtract the ones.	Subtract the tens.	Subtract the hundreds.
306 −89	2 10 3̸0̸6 −89	9 2 1̸0̸ 16 3̸0̸6̸ −89 ——— 7	9 2 1̸0̸ 16 3̸0̸6̸ −89 ——— 17	9 2 1̸0̸ 16 3̸0̸6̸ −89 ——— 217

Subtract.

	a	b	c	d	e	f
1.	154 −61	125 −74	107 −56	137 −50	124 −53	126 −71
2.	411 −80	429 −39	862 −91	428 −37	784 −93	605 −71
3.	984 −97	352 −65	604 −75	671 −93	325 −46	864 −79
4.	168 −69	906 −97	384 −96	763 −94	286 −98	857 −69
5.	666 −88	127 −49	700 −99	318 −79	722 −59	821 −48
6.	454 −67	357 −89	183 −94	306 −58	854 −95	347 −79
7.	346 −88	464 −87	207 −49	123 −34	253 −78	241 −59

Perfect score: 42   My score: _____

Problem Solving

Solve each problem.

1. A school building is 64 feet high. A flagpole is 125 feet high. How much higher is the flagpole?

The flag pole is _____ feet high.

The school building is _____ feet high.

The flag pole is _____ feet higher.

2. Steve threw a ball 117 feet. His cousin threw it 86 feet. How much farther did Steve throw it?

Steve threw the ball _____ feet.

His cousin threw the ball _____ feet.

Steve threw the ball _____ feet farther.

3. 865 tickets went on sale for a concert. So far 95 tickets have been sold. How many tickets are left?

_____ tickets went on sale.

_____ tickets have been sold.

_____ tickets are left.

4. Last year Mrs. Moore's bowling average was 91. This year her average is 123. How much has her bowling average improved over last year?

Her average has improved _____ points.

5. Trudy is reading a 234-page book. She has read 57 pages. How many more pages does she still have to read?

She has _____ pages yet to read.

1.

2.

3.

4.

5.

Perfect score: 11   My score: _____

22

Lesson 6   Addition and Subtraction

To check 62 + 57 = 119, subtract 57 from 119.

$$\begin{array}{r} 62 \\ +57 \\ \hline 119 \\ -57 \\ \hline 62 \end{array}$$ These should be the same.

To check 125 − 67 = 58, add 67 to 58.

$$\begin{array}{r} 125 \\ -67 \\ \hline 58 \\ +67 \\ \hline 125 \end{array}$$ These should be the same.

Add. Check each answer.

	a	b	c	d	e	f
1.	42 +25	63 +35	24 +64	26 +47	38 +42	59 +35
2.	63 +45	73 +56	83 +35	78 +25	62 +89	98 +95

Subtract. Check each answer.

	a	b	c	d	e	f
3.	85 −24	79 −45	68 −39	95 −28	68 −29	73 −48
4.	125 −63	146 −83	164 −73	104 −86	152 −64	186 −97

Perfect score: 24   My score: _____

Lesson 7   Addition and Subtraction

Add.

	a	b	c	d	e	f
1.	5 7 +9	3 5 +7	9 8 +6	5 3 +8	6 7 +5	4 3 +7
2.	38 +6	47 +9	63 +8	7 +49	8 +67	9 +47
3.	47 +24	43 +48	91 +26	46 +74	28 +356	435 +93
4.	15 22 +45	27 33 +18	31 12 +85	31 87 +41	17 156 +42	785 31 +56

Subtract.

	a	b	c	d	e	f
5.	36 −7	48 −9	63 −8	45 −7	54 −8	76 −9
6.	43 −25	75 −18	93 −44	50 −17	71 −28	67 −49
7.	147 −29	264 −25	784 −39	673 −54	270 −53	687 −59
8.	245 −93	705 −73	248 −75	638 −85	459 −87	317 −45
9.	345 −76	508 −59	867 −89	316 −39	707 −48	465 −68

Perfect score: 54   My score: _____

CHAPTER 2 TEST

Add. Check each answer.

	a	*b*	*c*	*d*	*e*	*f*
1.	3 7 +6	8 +4 7	5 9 +8	5 +4 8	6 3 +9	9 +8 1
2.	6 5 +2 7	3 4 +9 2	8 8 +3 7	1 5 9 +8 2	2 6 7 +7 6	3 4 7 +9 6

Subtract. Check each answer.

3.	8 4 −2 7	7 4 −4 5	6 8 −4 9	9 3 −3 8	6 1 −4 7	7 8 −5 9
4.	1 3 5 −6 4	1 2 6 −8 4	4 6 3 −7 2	1 5 3 −9 6	3 8 4 −9 6	3 0 2 −8 5

Solve.

5. Manny weighs 48 kilograms, Herm weighs 33 kilograms, and Alice weighs 37 kilograms. What is their combined weight?

5.

Their combined weight is _____ kilograms.

Perfect score: 25   My score: _____

PRE-TEST—Addition and Subtraction

Add.

	a	*b*	*c*	*d*	*e*	*f*
1.	621 +214	345 +332	426 +153	425 +316	245 +127	458 +329
2.	425 +193	373 +282	625 +193	624 +732	506 +792	591 +805
3.	397 +113	287 +125	926 +7287	3452 +1139	4646 +1283	5252 +3934
4.	31 10 +24	25 16 +23	232 151 +474	531 612 +743	2137 3272 +1324	4573 2281 +1654
5.	12 21 31 +12	200 413 134 +131	3171 1540 2134 +1023	3421 1313 1510 +2643	3117 2375 1132 +1214	3774 1571 3232 +1351

Subtract.

	a	*b*	*c*	*d*	*e*	*f*
6.	657 −234	745 −416	967 −173	406 −257	5627 −512	4848 −425
7.	4357 −138	5725 −273	6753 −902	7425 −286	8652 −937	6053 −782
8.	4357 −1132	5678 −1429	3675 −1294	5678 −2923	7802 −3254	9797 −1898
9.	67524 −1321	34723 −2308	78243 −4152	80145 −1913	76762 −9341	78545 −2837

Perfect score: 54   My score: _____

Lesson 1 Addition and Subtraction

Add the ones.	Add the tens.	Add the hundreds.	Subtract the ones.	Rename and subtract the tens.	Rename and subtract the hundreds.
783 +562 — 5	¹ 783 +562 — 45	¹ 783 + 562 — 1345	1253 −582 — 1	1 15 1253 −582 — 71	11 1̸1̸15 1253 − 582 — 671

Add.

	a	b	c	d	e	f
1.	432 +325	325 +536	232 +573	325 +814	328 +193	529 +287
2.	675 +907	867 +325	625 +594	891 +536	675 +738	475 +969
3.	357 +528	146 +494	734 +859	536 +673	867 +795	893 +757

Subtract.

	a	b	c	d	e
4.	857 −143	757 −129	467 −182	952 −278	863 −389
5.	1489 −527	1246 −813	1578 −927	1728 −919	1373 −548
6.	1458 −773	1732 −961	1420 −735	1254 −995	1652 −797
7.	1020 −552	1234 − 435	1600 − 900	1357 − 968	1000 − 879

Perfect score: **38**   My score: _____

Problem Solving

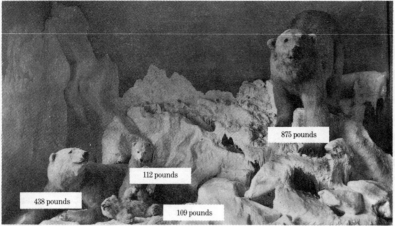

875 pounds

112 pounds

438 pounds

109 pounds

Solve each problem.

1. Find the combined weight of the two smallest polar bears.

Are you to add or subtract? _____

The combined weight is _____ pounds.

1.

2. How much more does the largest bear weigh than the smallest bear?

Are you to add or subtract? _____

The largest bear weighs _____ pounds more than the smallest bear.

2.

3. The adult male bear is 118 inches long. The adult female bear is 69 inches long. How much longer is the male than the female?

Are you to add or subtract? _____

The male bear is _____ inches longer than the female bear.

3.

4. Find the combined weight of the two largest polar bears.

Are you to add or subtract? _____

The combined weight is _____ pounds.

4.

Perfect score: 8   My score: _____

28

Lesson 2 Addition and Subtraction

NAME _____

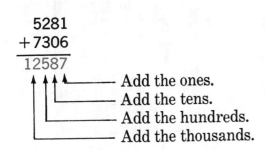

Add the ones.
Add the tens.
Add the hundreds.
Add the thousands.

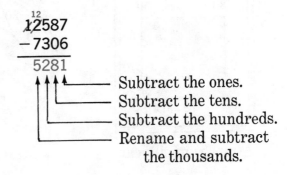

Subtract the ones.
Subtract the tens.
Subtract the hundreds.
Rename and subtract
the thousands.

Add.

	a	*b*	*c*	*d*	*e*
1.	3253 +1424	6725 +1138	2456 +3293	5263 +1824	6245 +7403
2.	5478 +2325	5769 +3914	6518 +9246	5382 +1664	7683 +8185
3.	8121 +7934	2523 +4694	6378 +7164	3235 +4917	8483 +9658

Subtract.

	a	*b*	*c*	*d*	*e*
4.	2675 −436	3675 −594	6725 −904	4307 −289	6352 −764
5.	4357 −2349	6735 −1264	9525 −4603	5675 −2389	6052 −2968
6.	16784 −7325	12543 −3362	15747 −6936	10137 −9652	17675 −8896
7.	14321 −3421	10236 −1527	13500 −6870	11238 −4739	10000 −6782

Perfect score: 35   My score: _____

29

Problem Solving

Mrs. Crannell's car

Mr. Comm's car

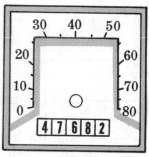

Mr. Hsu's car

Solve each problem.

1. Mr. Hsu's car has been driven how many more miles than Mr. Comm's car?

Mr. Hsu's car has been driven _____ miles.

Mr. Comm's car has been driven _____ miles.

Mr. Hsu's car has been driven _____ miles more.

2. Mr. Hsu plans to drive his car 870 miles on a trip. How many miles will be on his car after the trip?

_____ miles are on his car now.

The trip will be _____ miles long.

_____ miles will be on his car after the trip.

3. Mrs. Crannell's car needs an oil change at 6,000 miles. How many more miles can she go before changing oil?

She can go _____ more miles.

4. Mr. Mitchel wants to buy a station wagon for $8,365 or a sedan for $6,978. How much more does the station wagon cost?

The station wagon costs $_____ more.

5. St. Louis is 1,933 kilometers from Boston. San Francisco is 3,414 kilometers from St. Louis. How far is it from Boston to San Francisco by way of St. Louis?

The distance is _____ kilometers.

1.	
2.	**3.**
4.	**5.**

Perfect score: 9   My score: _____

30

Lesson 3 Addition

Add in each place-value position from right to left.

```
               1 1          2 1 2        1 2 2
              6374          1632         4325
            + 2809          3779         6078
            ------         3779          5298
              9183       + 6809        + 5764
                          ------        ------
                          12220         21465
```

Add.

	a	b	c	d	e
1.	24 31 +40	46 23 +15	45 62 +71	45 13 +21	52 23 +71
2.	34 21 +112	126 12 +624	345 162 +71	524 630 +721	305 131 +422
3.	3235 3112 +1486	2145 3418 +1932	8218 3245 +4123	1353 2331 +3642	4435 8271 +4160
4.	5641 2722 +4833	1826 2574 +4493	7137 8028 +7656	2453 8742 +2561	3417 8703 +2854
5.	5247 2403 1125 +1017	3253 1161 1172 +4080	1601 2722 3813 +1241	5145 6201 2312 +4021	1011 2462 3571 +1254
6.	1025 3113 1258 +2464	1546 2335 3822 +1941	4124 1231 5352 +6075	3652 6274 3175 +5112	6317 2164 5573 +4258

Perfect score: 30   My score: _____

Problem Solving

Cab number	Liters of gasoline used	Number of hours driven	Number of kilometers traveled	Number of passengers	Amount of fares collected
100	707	566	6,307	2,205	$8,221
101	791	573	6,962	2,542	$8,825
102	729	548	6,566	1,821	$8,533

Dot Cab Company
May Report

Solve each problem.

1. How many liters of gasoline were used by the three cabs during the month of May?

_____ liters were used by cab 100.

_____ liters were used by cab 101.

_____ liters were used by cab 102.

_____ liters were used in all.

2. How many hours were the three cabs driven?

The cabs were driven _____ hours.

3. How many passengers rode in the three cabs?

_____ passengers rode in the cabs.

4. How many kilometers did the three cabs travel?

The cabs traveled _____ kilometers.

5. How much was collected in fares for the three cabs?

$_____ was collected in fares.

1.

2.

3.

4.

5.

Perfect score: 8   My score: _____

32

Lesson 4 Addition and Subtraction

Add.

	a	b	c	d	e
1.	312 +541	135 +427	231 +384	532 +614	537 +148
2.	1456 +218	295 +2461	723 +5413	4561 +349	4624 +597
3.	5247 +2216	4673 +4285	4672 +1401	5314 +7053	4314 +3197
4.	42 13 +25	131 12 +449	213 227 +384	4163 5274 +6521	4712 1824 +5531
5.	67 78 55 +27	731 142 253 +461	7325 2106 7347 +2511	5314 6024 7151 +2235	5678 2345 6789 +4257

Subtract.

	a	b	c	d	e
6.	687 −434	754 −236	576 −393	605 −388	795 −498
7.	1234 −125	3857 −665	4257 −843	4657 −839	5014 −968
8.	7354 −4038	5619 −2348	4187 −2574	6753 −1942	7815 −4176
9.	42573 −1846	36154 −9038	46124 −9762	54751 −2896	70534 −7689

Perfect score: 45   My score: _____

33

Problem Solving

Solve each problem.

1. An empty truck weighs 1,750 kilograms. It is to be loaded with 1,402 kilograms of cargo. What will be the weight of the truck and its cargo?

The truck weighs _____ kilograms.

The cargo weighs _____ kilograms.

The combined weight is _____ kilograms.

2. 9,852 fans attended the game on Monday. There were 7,569 fans at the game on Tuesday. How many fans attended the two games?

_____ fans attended the games.

3. Ms. Krom has saved $10,320 for a new car. Her goal is to save $12,500. How much more must she save?

She must save $_____ more.

4. Complete the table.

5. How many more women are there than men?

There are

_____ more women.

Benny School	
Number of women	1,029
Number of men	983
Total	

6. Complete the table.

7. How many more votes did Witt get than Lewis?

Witt got

_____ more votes.

Election results	Number of votes
Witt	4,327
Lewis	2,539
Total	

1.

2.

3.

5.

7.

Perfect score: 9   My score: _____

34

CHAPTER 3 TEST

Add.

	a	*b*	*c*	*d*	*e*
1.	342 +325	725 +146	362 +475	425 +1723	284 +1523
2.	3156 +1327	2363 +4195	3741 +2625	8403 +3445	3456 +2157
3.	43 17 +13	435 16 +127	4113 1590 +2671	234 357 214 +526	5253 4376 2416 +1327

Subtract.

	a	*b*	*c*	*d*
4.	625 −407	908 −436	1765 −934	2576 −882
5.	5724 −1543	6753 −1908	17024 −9653	67543 −9988

Solve each problem.

6. The Sears Tower is 1,454 feet tall. It has a 346-foot TV antenna on top. What is the height of the building and antenna?

6.

The height is _____ feet.

7. A space rocket is going 9,785 kilometers per hour. Later it will go 23,650 kilometers per hour. How much will the speed have to increase?

7.

The speed will increase _____ kilometers per hour.

Perfect score: 25   My score: _____

35

PRE-TEST—Multiplication

Multiply.

	a	b	c	d	e	f
1.	7 ×4	6 ×9	8 ×5	40 ×2	20 ×3	10 ×8
2.	23 ×2	21 ×4	32 ×3	11 ×5	21 ×3	22 ×2
3.	22 ×3	21 ×2	43 ×2	23 ×3	22 ×4	34 ×2
4.	17 ×5	27 ×3	15 ×6	23 ×4	12 ×7	12 ×8
5.	51 ×9	62 ×4	71 ×8	67 ×7	96 ×3	83 ×5
6.	200 ×4	300 ×2	400 ×2	121 ×4	124 ×2	312 ×3
7.	105 ×9	124 ×4	325 ×3	121 ×8	172 ×4	283 ×3
8.	400 ×7	412 ×3	924 ×4	513 ×7	618 ×4	176 ×5
9.	830 ×7	731 ×6	138 ×7	673 ×8	469 ×5	869 ×4

Perfect score: 54   My score: _____

Lesson 1 Multiplication

NAME _____

5 → Find the ⬚5⬚ -row.

×6 → Find the ⬚6⬚ -column.

30 ← The product is named where the 5-row and 6-column meet.

Use the table to multiply.

$$\begin{array}{r} 7 \\ \times 9 \\ \hline \end{array}$$

6-column

×	0	1	2	3	4	5	6	7	8	9
0	0	0	0	0	0	0	0	0	0	0
1	0	1	2	3	4	5	6	7	8	9
2	0	2	4	6	8	10	12	14	16	18
3	0	3	6	9	12	15	18	21	24	27
4	0	4	8	12	16	20	24	28	32	36
5	0	5	10	15	20	25	30	35	40	45
6	0	6	12	18	24	30	36	42	48	54
7	0	7	14	21	28	35	42	49	56	63
8	0	8	16	24	32	40	48	56	64	72
9	0	9	18	27	36	45	54	63	72	81

5-row →

Multiply.

	a	b	c	d	e	f	g	h
1.	6 ×1	7 ×8	8 ×9	9 ×0	7 ×7	6 ×7	7 ×1	4 ×6
2.	9 ×9	6 ×2	7 ×6	9 ×1	8 ×1	6 ×8	8 ×8	3 ×7
3.	8 ×0	7 ×5	6 ×0	9 ×2	8 ×7	6 ×9	7 ×0	4 ×7
4.	5 ×6	9 ×3	7 ×4	6 ×3	9 ×4	4 ×8	8 ×6	3 ×6
5.	7 ×9	5 ×7	5 ×8	8 ×5	6 ×4	7 ×3	9 ×5	2 ×6
6.	7 ×2	8 ×4	9 ×6	9 ×7	4 ×9	6 ×5	9 ×8	2 ×8
7.	8 ×3	5 ×9	3 ×8	8 ×2	3 ×9	2 ×7	6 ×6	2 ×9

Perfect score: 56   My score: _____

37

Problem Solving

Solve each problem.

1. Last week Tommy's father worked five 8-hour shifts. How many hours did he work last week?

He worked _____ shifts.

There were _____ hours in each shift.

He worked _____ hours last week.

2. A certain factory operates two 8-hour shifts each day. How many hours does the factory operate each day?

There are _____ shifts.

There are _____ hours in each shift.

The factory operates _____ hours each day.

3. It takes the clean-up crew 4 hours to clean the factory after each day's work. How many hours will the clean-up crew work during a 5-day week?

The clean-up crew works _____ hours a day.

They work _____ days a week.

The clean-up crew works _____ hours a week.

4. Gloria's mother works 5 hours each day. She works 5 days each week. How many hours does she work each week?

She works _____ hours each week.

5. It costs a company $8 an hour to operate a certain machine. How much will it cost to operate the machine for 6 hours?

It will cost $_____.

1.

2.

3.

4.   **5.**

Perfect score: 11   My score: _____

38

Lesson 2   Multiplication

$$\begin{array}{r} 4 \\ \times 2 \\ \hline 8 \end{array} \qquad \begin{array}{r} 40 \\ \times 2 \\ \hline 80 \end{array}$$

If $2 \times 4 = 8$,
then $2 \times 40 = 80$.

	Multiply 3 ones by 2.	Multiply 4 tens by 2.
$\begin{array}{r} 43 \\ \times 2 \end{array}$	$\begin{array}{r} 43 \\ \times 2 \\ \hline 6 \end{array}$	$\begin{array}{r} 43 \\ \times 2 \\ \hline 86 \end{array}$

Multiply.

	a	b	c	d	e	f
1.	$\begin{array}{r} 3 \\ \times 3 \end{array}$	$\begin{array}{r} 30 \\ \times 3 \end{array}$	$\begin{array}{r} 2 \\ \times 4 \end{array}$	$\begin{array}{r} 20 \\ \times 4 \end{array}$	$\begin{array}{r} 3 \\ \times 2 \end{array}$	$\begin{array}{r} 30 \\ \times 2 \end{array}$
2.	$\begin{array}{r} 2 \\ \times 3 \end{array}$	$\begin{array}{r} 30 \\ \times 3 \end{array}$	$\begin{array}{r} 32 \\ \times 3 \end{array}$	$\begin{array}{r} 1 \\ \times 2 \end{array}$	$\begin{array}{r} 40 \\ \times 2 \end{array}$	$\begin{array}{r} 41 \\ \times 2 \end{array}$
3.	$\begin{array}{r} 11 \\ \times 9 \end{array}$	$\begin{array}{r} 33 \\ \times 3 \end{array}$	$\begin{array}{r} 12 \\ \times 3 \end{array}$	$\begin{array}{r} 14 \\ \times 2 \end{array}$	$\begin{array}{r} 31 \\ \times 3 \end{array}$	$\begin{array}{r} 13 \\ \times 3 \end{array}$
4.	$\begin{array}{r} 32 \\ \times 2 \end{array}$	$\begin{array}{r} 23 \\ \times 3 \end{array}$	$\begin{array}{r} 42 \\ \times 2 \end{array}$	$\begin{array}{r} 21 \\ \times 4 \end{array}$	$\begin{array}{r} 13 \\ \times 2 \end{array}$	$\begin{array}{r} 11 \\ \times 6 \end{array}$
5.	$\begin{array}{r} 12 \\ \times 2 \end{array}$	$\begin{array}{r} 11 \\ \times 5 \end{array}$	$\begin{array}{r} 33 \\ \times 2 \end{array}$	$\begin{array}{r} 11 \\ \times 3 \end{array}$	$\begin{array}{r} 21 \\ \times 2 \end{array}$	$\begin{array}{r} 22 \\ \times 3 \end{array}$
6.	$\begin{array}{r} 11 \\ \times 4 \end{array}$	$\begin{array}{r} 44 \\ \times 2 \end{array}$	$\begin{array}{r} 22 \\ \times 2 \end{array}$	$\begin{array}{r} 11 \\ \times 8 \end{array}$	$\begin{array}{r} 11 \\ \times 2 \end{array}$	$\begin{array}{r} 13 \\ \times 2 \end{array}$
7.	$\begin{array}{r} 23 \\ \times 2 \end{array}$	$\begin{array}{r} 22 \\ \times 4 \end{array}$	$\begin{array}{r} 24 \\ \times 2 \end{array}$	$\begin{array}{r} 21 \\ \times 3 \end{array}$	$\begin{array}{r} 31 \\ \times 2 \end{array}$	$\begin{array}{r} 11 \\ \times 7 \end{array}$

Perfect score: 42   My score: _____

Problem Solving

Solve each problem.

1. There are 1 dozen cans of peaches in each carton. How many cans are in 2 cartons? Remember, there are 12 items in 1 dozen.

_____ cans are in 1 dozen.

There are _____ cartons.

There are _____ cans of peaches in 2 cartons.

2. 12 cans of pineapple are in each carton. How many cans are in 3 cartons?

There are _____ cans of pineapple in 3 cartons.

3. There are 1 dozen cans of pears in each carton. How many cans are in 4 cartons?

There are _____ cans of pears in 4 cartons.

4. Ten cans of orange sections come in each carton. How many cans are in 4 cartons?

There are _____ cans of orange sections in 4 cartons.

1.	2.
3.	4.

Perfect score: 6   My score: _____

40

Lesson 3   Multiplication

Multiply
8 ones by 3.

$$3 \times 8 = 24 \text{ or } 20 + 4$$

Multiply 7 tens by 3.
Add the 2 tens.

$$3 \times 70 = 210$$
$$210 + 20 = 230 \text{ or } 200 + 30$$

```
78        78         78
×3        ×3         ×3
          ‾4        234
```

Multiply.

	a	b	c	d	e	f
1.	28 ×2	23 ×4	25 ×3	16 ×6	19 ×5	37 ×2
2.	14 ×6	13 ×7	29 ×3	12 ×8	46 ×2	12 ×7
3.	53 ×3	61 ×5	74 ×2	81 ×6	71 ×7	62 ×4
4.	92 ×4	73 ×2	91 ×5	61 ×7	72 ×3	61 ×8
5.	73 ×9	85 ×2	59 ×6	48 ×7	67 ×3	57 ×8
6.	72 ×6	76 ×4	83 ×9	98 ×5	54 ×7	42 ×9
7.	83 ×5	74 ×6	97 ×9	58 ×8	74 ×7	49 ×4

Perfect score: 42   My score: _____

Problem Solving

Solve each problem.

1. A bottle of Sudsy Shampoo contains 12 fluid ounces. How many fluid ounces are in 3 such bottles?

There are _____ fluid ounces in each bottle.

There are _____ bottles.

There are _____ fluid ounces in 3 bottles.

2. Mr. Long drives 14 miles to work each day. He works 6 days a week. How far does he drive to work each week?

He drives _____ miles each day.

He works _____ days each week.

He drives _____ miles each week.

3. A case contains 24 cans. How many cans will be in 9 such cases?

Each case contains _____ cans.

There are _____ cases.

There are _____ cans in 9 cases.

4. The Acme Salt Company shipped 8 sacks of salt to the Sour Pickle Company. Each sack of salt weighed 72 pounds. What was the weight of the shipment?

The total weight of the shipment was _____ pounds.

5. A train can travel 69 kilometers in one hour. How far can it travel in 4 hours?

It can travel _____ kilometers.

1.	
2.	
3.	
4.	**5.**

Perfect score: 11   My score: _____

42

Lesson 4 Multiplication

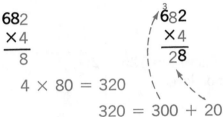

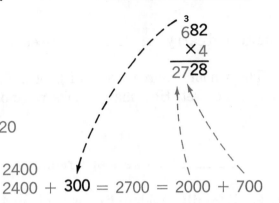

```
  4    400
 ×6    ×6
 ──    ────
 24   2400
```

Multiply
2 ones by 4.

```
 682
  ×4
 ───
   8
```

Multiply
8 tens by 4.

```
  ³
 682
  ×4
 ───
  28
```

$4 \times 80 = 320$

$320 = 300 + 20$

Multiply 6 hundreds by 4.
Add the 3 hundreds.

```
  ³
 682
  ×4
 ────
 2728
```

$4 \times 600 = 2400$

$2400 + \mathbf{300} = 2700 = 2000 + 700$

Multiply.

	a	b	c	d	e	f
1.	3 ×2	300 ×2	5 ×7	500 ×7	6 ×8	600 ×8
2.	18 ×4	200 ×4	218 ×4	21 ×4	700 ×4	721 ×4

Multiply.

	a	b	c	d	e
3.	321 ×3	423 ×2	212 ×4	349 ×2	327 ×3
4.	156 ×6	238 ×4	807 ×6	413 ×7	421 ×8
5.	987 ×6	478 ×9	678 ×7	727 ×8	594 ×5

Perfect score: 27   My score: _____

Problem Solving

Solve each problem.

1. An airplane can carry 183 passengers. How many passengers could 5 such airplanes carry?

They could carry _____ passengers.

2. The school ordered 7 sets of books. There are 125 books in each set. How many books were ordered?

_____ books were ordered.

3. Each family in a building was given 4 keys. There were 263 families in the building. How many keys were given out?

_____ keys were given out.

4. Each pupil receives 2 cartons of milk a day. There are 912 pupils in the school. How many cartons of milk will be needed?

_____ cartons of milk will be needed.

5. There are 217 apartments in Jane's building. Each apartment has 6 windows. How many windows are there in all?

There are _____ windows in all.

6. Elmer the elephant eats 145 pounds of food a day. How many pounds of food will he eat in 7 days?

Elmer will eat _____ pounds of food in 7 days.

7. Carlos delivers 128 papers each day. How many papers will he deliver in 6 days?

He will deliver _____ papers in 6 days.

1.

2.

3.

4.

5.

6.

7.

Perfect score: 7   My score: _____

44

Lesson 5 Multiplication

Multiply.

	a	b	c	d	e	f
1.	1 1 ×7	3 2 ×2	2 1 ×4	3 2 ×3	2 4 ×2	2 3 ×3
2.	1 3 ×7	1 5 ×5	4 7 ×2	1 3 ×6	2 4 ×4	2 5 ×3
3.	7 2 ×3	9 2 ×4	7 1 ×6	6 2 ×4	8 1 ×5	9 3 ×3
4.	6 6 ×4	4 8 ×7	6 7 ×6	7 2 ×9	5 7 ×7	4 9 ×5

Multiply.

	a	b	c	d	e
5.	2 1 4 ×2	2 3 1 ×3	2 1 0 ×4	2 2 4 ×4	1 1 5 ×6
6.	1 1 7 ×5	1 3 1 ×7	3 9 4 ×2	1 2 1 ×8	7 3 2 ×3
7.	9 1 2 ×4	6 1 0 ×5	1 5 7 ×6	2 3 5 ×4	1 3 7 ×7
8.	4 0 6 ×8	6 1 3 ×7	5 2 7 ×3	6 7 1 ×9	4 3 0 ×8
9.	9 4 1 ×7	4 8 3 ×8	6 7 5 ×9	7 2 9 ×8	4 3 6 ×7

Perfect score: 49   My score: _____

Problem Solving

Solve each problem.

1. Each side of a baseball diamond is 90 feet in length. How far is it around the baseball diamond?

It is _____ feet around the diamond.

2. Lois practices her flute 35 minutes each day. How many minutes does she practice during a week?

She practices _____ minutes each week.

3. A certain light fixture contains six 75-watt light bulbs. When all the bulbs are on, how many watts are being used?

The 6 bulbs are using _____ watts.

4. A bus fare is 45 cents. What is the cost of 2 fares?

The cost is _____ cents.

5. An aircraft carrier is as long as 3 football fields. A football field is 360 feet long. How long is the aircraft carrier?

It is _____ feet long.

6. A shipment has 5 boxes. Each box weighs 125 kilograms. What is the weight of the shipment?

The weight is _____ kilograms.

7. An airplane can travel 635 kilometers in one hour. How far can it travel in 4 hours?

It can travel _____ kilometers in 4 hours.

8. A certain machine can produce 265 items an hour. At this rate, how many items can be produced in 8 hours?

_____ items can be produced.

1.	2.
3.	**4.**
5.	**6.**
7.	**8.**

Perfect score: 8   My score: _____

CHAPTER 4 TEST

Multiply.

	a	b	c	d	e	f
1.	30 ×2	20 ×3	10 ×5	40 ×2	30 ×3	10 ×9
2.	32 ×3	43 ×2	21 ×4	23 ×3	11 ×6	12 ×4
3.	12 ×7	19 ×4	28 ×3	16 ×5	46 ×2	12 ×8
4.	60 ×7	80 ×9	40 ×7	71 ×6	92 ×4	81 ×6
5.	72 ×8	63 ×9	57 ×6	39 ×7	49 ×5	65 ×4

Multiply.

	a	b	c	d	e
6.	432 ×2	112 ×4	216 ×4	113 ×5	103 ×7
7.	131 ×7	282 ×3	711 ×5	612 ×4	932 ×3
8.	137 ×7	124 ×8	513 ×7	306 ×9	417 ×4
9.	521 ×9	941 ×8	567 ×4	439 ×5	857 ×6

Perfect score: 50   My score: _____

NAME _____

Multiply.

	a	b	c	d	e	f
1.	32 ×20	23 ×30	11 ×70	64 ×40	49 ×80	27 ×90
2.	43 ×21	23 ×32	34 ×22	23 ×39	21 ×48	32 ×37
3.	37 ×61	53 ×41	38 ×82	74 ×56	47 ×68	76 ×54

Multiply.

	a	b	c	d	e
4.	402 ×21	320 ×32	312 ×38	524 ×17	423 ×29
5.	624 ×61	213 ×53	431 ×82	426 ×57	834 ×68

Perfect score: 28   My score: _____

Lesson 1   Multiplication

$$\begin{array}{r} 21 \\ \times 34 \\ \hline \end{array}$$

Multiply 21 by 4 ones.

$$\begin{array}{r} 21 \\ \times 4 \\ \hline 84 \end{array}$$

Multiply 21 by 3 tens.

$$\begin{array}{r} 21 \\ \times 34 \\ \hline 84 \\ 630 \end{array}$$

$$\left.\begin{array}{r} 21 \\ \times 34 \\ \hline 84 \\ 630 \\ \hline 714 \end{array}\right\} \text{Add.}$$

Multiply.

	a	*b*	*c*	*d*	*e*	*f*
1.	$\begin{array}{r}13\\\times 3\\\hline\end{array}$	$\begin{array}{r}13\\\times 30\\\hline\end{array}$	$\begin{array}{r}43\\\times 2\\\hline\end{array}$	$\begin{array}{r}43\\\times 20\\\hline\end{array}$	$\begin{array}{r}42\\\times 20\\\hline\end{array}$	$\begin{array}{r}23\\\times 30\\\hline\end{array}$
2.	$\begin{array}{r}34\\\times 2\\\hline\end{array}$	$\begin{array}{r}34\\\times 10\\\hline\end{array}$	$\begin{array}{r}34\\\times 12\\\hline\end{array}$	$\begin{array}{r}32\\\times 3\\\hline\end{array}$	$\begin{array}{r}32\\\times 20\\\hline\end{array}$	$\begin{array}{r}32\\\times 23\\\hline\end{array}$
3.	$\begin{array}{r}21\\\times 42\\\hline\end{array}$	$\begin{array}{r}33\\\times 32\\\hline\end{array}$	$\begin{array}{r}79\\\times 11\\\hline\end{array}$	$\begin{array}{r}31\\\times 31\\\hline\end{array}$	$\begin{array}{r}11\\\times 27\\\hline\end{array}$	$\begin{array}{r}12\\\times 32\\\hline\end{array}$
4.	$\begin{array}{r}33\\\times 23\\\hline\end{array}$	$\begin{array}{r}22\\\times 43\\\hline\end{array}$	$\begin{array}{r}56\\\times 11\\\hline\end{array}$	$\begin{array}{r}23\\\times 32\\\hline\end{array}$	$\begin{array}{r}14\\\times 21\\\hline\end{array}$	$\begin{array}{r}11\\\times 94\\\hline\end{array}$

Perfect score: 24   My score: _____

Problem Solving

Solve each problem.

1. Each train car can carry 10 vans. There are 21 train cars full of vans. How many vans are there in all?

Each train car can carry _____ vans.

There are _____ train cars carrying vans.

There are _____ vans in all.

1.

2. Each train car can carry 15 automobiles. There are 32 train cars full of automobiles. How many automobiles are there in all?

Each train car can carry _____ automobiles.

There are _____ train cars carrying automobiles.

There are _____ automobiles in all.

2.

3. Suppose there had been 48 train cars in Problem 2. Then how many automobiles are there in all?

There are _____ automobiles in all.

3.

Perfect score: 7   My score: _____

50

Lesson 2 Multiplication

	Multiply 47 by 2 ones.		Multiply 47 by 3 tens.	

$$\begin{array}{r} 47 \\ \times 32 \\ \hline \end{array}$$

$$\begin{array}{r} 47 \\ \times 32 \\ \hline 94 \end{array}$$

$$\begin{array}{r} 47 \\ \times 32 \\ \hline 94 \\ 1410 \end{array}$$

$$\left.\begin{array}{r} 47 \\ \times 32 \\ \hline 94 \\ 1410 \\ \hline 1504 \end{array}\right\} \text{Add.}$$

Multiply.

	a	b	c	d	e	f
1.	$\begin{array}{r}65\\\times7\\\hline\end{array}$	$\begin{array}{r}65\\\times70\\\hline\end{array}$	$\begin{array}{r}37\\\times8\\\hline\end{array}$	$\begin{array}{r}37\\\times80\\\hline\end{array}$	$\begin{array}{r}64\\\times40\\\hline\end{array}$	$\begin{array}{r}75\\\times50\\\hline\end{array}$
2.	$\begin{array}{r}57\\\times19\\\hline\end{array}$	$\begin{array}{r}31\\\times37\\\hline\end{array}$	$\begin{array}{r}72\\\times16\\\hline\end{array}$	$\begin{array}{r}43\\\times23\\\hline\end{array}$	$\begin{array}{r}42\\\times29\\\hline\end{array}$	$\begin{array}{r}39\\\times17\\\hline\end{array}$
3.	$\begin{array}{r}83\\\times91\\\hline\end{array}$	$\begin{array}{r}43\\\times62\\\hline\end{array}$	$\begin{array}{r}44\\\times52\\\hline\end{array}$	$\begin{array}{r}32\\\times63\\\hline\end{array}$	$\begin{array}{r}34\\\times32\\\hline\end{array}$	$\begin{array}{r}52\\\times41\\\hline\end{array}$
4.	$\begin{array}{r}76\\\times49\\\hline\end{array}$	$\begin{array}{r}97\\\times94\\\hline\end{array}$	$\begin{array}{r}63\\\times85\\\hline\end{array}$	$\begin{array}{r}47\\\times65\\\hline\end{array}$	$\begin{array}{r}83\\\times78\\\hline\end{array}$	$\begin{array}{r}92\\\times86\\\hline\end{array}$

Perfect score: 24   My score: _____

Problem Solving

Solve each problem.

1. A school bus can carry 66 pupils. How many pupils can ride on 12 such buses?

_____ pupils can ride on 12 buses.

2. A train can travel 97 kilometers in one hour. How far can it travel in 13 hours?

It can travel _____ kilometers.

3. A copy machine can make 45 copies per minute. How many copies can it make in 15 minutes?

It can make _____ copies in 15 minutes.

4. Ira bought 12 pizza pies for a party. Each pizza was cut into 16 pieces. How many pieces did he have in all?

He had _____ pieces in all.

5. Miss Lens bought 18 rolls of film. Thirty-six pictures can be taken on each roll. How many pictures can she take?

_____ pictures can be taken.

6. Mr. Dzak works 40 hours each week. How many hours will he work in 26 weeks?

He will work _____ hours.

7. A car is traveling 23 meters per second. How far will the car travel in 50 seconds?

It will travel _____ meters.

8. There are 24 hours in a day. How many hours are there in 49 days?

There are _____ hours in 49 days.

1.	2.
3.	**4.**
5.	**6.**
7.	**8.**

Perfect score: 8   My score: _____

52

Lesson 3   Multiplication

Multiply.

	a	*b*	*c*	*d*	*e*	*f*
1.	24 ×20	32 ×30	21 ×40	56 ×20	75 ×60	84 ×70
2.	42 ×21	32 ×23	22 ×42	23 ×13	33 ×23	24 ×22
3.	43 ×25	87 ×17	34 ×25	32 ×28	21 ×48	32 ×39
4.	54 ×71	43 ×72	32 ×63	42 ×82	83 ×51	34 ×92
5.	68 ×73	42 ×58	49 ×86	37 ×94	62 ×48	28 ×59

Perfect score: 30   My score: _____

Problem Solving

Solve each problem.

1. Mrs. Carter can type 55 words a minute. How many words can she type in 15 minutes?

She can type _____ words.

2. A machine puts caps on bottles at a rate of 96 per minute. At that rate, how many bottles can be capped in 25 minutes?

_____ bottles can be capped in 25 minutes.

3. Mr. Oliver travels 28 kilometers getting to and from work each day. How many kilometers will he travel in 22 working days?

He will travel _____ kilometers.

4. There are 48 thumbtacks in a box. How many are there in 15 boxes?

There are _____ thumbtacks.

5. There are 24 cars on a train. Suppose 66 passengers can ride in each car. How many passengers can ride in the train?

_____ passengers can ride in the train.

6. Carlos delivers 75 papers each day. How many papers will he deliver in 14 days?

He will deliver _____ papers.

7. A new building is to be 16 stories high. There are to be 14 feet for each story. How high will the building be?

The building will be _____ feet high.

1.	

2.	**3.**

4.	**5.**

6.	**7.**

Perfect score: 7   My score: _____

54

Lesson 4 Multiplication

	Multiply 512 by 3 ones.	Multiply 512 by 2 tens.	

```
  512          512          512          512
 ×23          ×23          ×23          ×23
             ────         ────         ────
             1536         1536         1536 ⎫ Add.
                         10240        10240 ⎭
                                      ─────
                                      11776
```

Multiply.

	a	b	c	d	e	f
1.	615 ×3	615 ×30	728 ×4	728 ×40	555 ×60	783 ×50
2.	132 ×4	132 ×20	132 ×24	323 ×3	323 ×60	323 ×63

Multiply.

	a	b	c	d	e
3.	212 ×23	423 ×12	121 ×49	321 ×37	412 ×24
4.	324 ×82	343 ×62	429 ×63	749 ×96	476 ×83

Perfect score: 22   My score: _____

Problem Solving

Solve each problem.

1. Gertrude plans to swim 150 laps this week. Each lap is 50 meters. How many meters does she plan to swim?

She plans to swim _____ meters.

2. Each hour 225 pictures can be developed. How many pictures can be developed in 12 hours?

_____ pictures can be developed.

3. One section of a sports arena has 24 rows of seats. There are 125 seats in each row. How many seats are there in that section?

There are _____ seats in that section.

4. There are 144 bags of salt in a shipment. Each bag weighs 36 kilograms. What is the weight of the shipment?

The weight is _____ kilograms.

5. James sells 165 papers a day. How many papers will he sell in 28 days?

He will sell _____ papers.

6. A jet cruises at 575 miles an hour. At that rate, how many miles will it travel in 12 hours?

It will travel _____ miles.

7. A certain desk weighs 118 kilograms. How many kilograms would 15 of the desks weigh?

They would weigh _____ kilograms.

8. 23 workers will deliver 290 circulars each. How many circulars will be delivered in all?

_____ circulars will be delivered in all.

1.	**2.**
3.	**4.**
5.	**6.**
7.	**8.**

Perfect score: 8   My score: _____

56

Lesson 5 Multiplication

Multiply.

	a	b	c	d	e	f
1.	54 ×21	75 ×42	63 ×39	27 ×64	84 ×56	67 ×93
2.	69 ×17	87 ×65	49 ×78	62 ×89	39 ×47	78 ×36

Multiply.

	a	b	c	d	e
3.	304 ×57	540 ×63	327 ×48	428 ×76	548 ×91
4.	924 ×27	286 ×14	478 ×82	375 ×89	721 ×52
5.	131 ×27	937 ×83	205 ×74	240 ×96	121 ×57

Perfect score: 27   My score: _____

Problem Solving

Solve each problem.

1. There are 24 slices of bread in a loaf. How many slices are there in 25 loaves?

There are _____ slices.

2. There are 500 sheets in a giant pack of notebook paper. How many sheets are there in 12 giant packs?

There are _____ sheets.

3. There are 12 eggs in a dozen. How many eggs are there in 16 dozen?

There are _____ eggs in 16 dozen.

4. There are 180 eggs packed in a case. How many eggs are there in 24 cases?

There are _____ eggs.

5. Fifty stamps are needed to fill each page of a stamp book. The book contains 24 pages. How many stamps are needed to fill the book?

_____ stamps are needed.

6. Marsha practices the piano 35 minutes each day. How many minutes will she practice in 28 days?

She will practice _____ minutes.

7. The average weight of the 11 starting players on a football team is 79 kilograms. What is the combined weight of the 11 players?

_____ kilograms is the combined weight.

8. There are 328 pages in each of the 16 volumes of an encyclopedia. How many pages are there in all?

There are _____ pages in all.

1.	2.
3.	4.
5.	6.
7.	8.

Perfect score: 8   My score: _____

CHAPTER 5 TEST

Multiply.

	a	*b*	*c*	*d*	*e*
1.	21 ×43	23 ×13	34 ×21	34 ×25	23 ×36
2.	63 ×51	42 ×72	23 ×63	72 ×89	56 ×39
3.	314 ×21	123 ×32	302 ×23	527 ×15	607 ×41
4.	324 ×26	342 ×82	312 ×37	321 ×73	423 ×29
5.	682 ×57	428 ×96	357 ×85	537 ×49	729 ×84

5

Perfect score: 25   My score: _____

PRE-TEST—Multiplication

Multiply.

	a	b	c	d
1.	5000 ×7	2107 ×4	3251 ×3	4731 ×2
2.	7131 ×5	7652 ×8	2121 ×24	6742 ×17
3.	4132 ×62	8767 ×71	5264 ×69	4675 ×78
4.	321 ×300	567 ×400	312 ×320	432 ×207
5.	423 ×912	413 ×792	729 ×816	875 ×438

Perfect score: 20   My score: _____

Lesson 1 Multiplication

NAME _____

		Multiply 2 ones by 4.	Multiply 3 tens by 4.	Multiply 2 hundreds by 4. Add the 1 hundred.	Multiply 6 thousands by 4.
6 ×4 24	6000 ×4 24000	6232 ×4 8	$6\overset{1}{2}32$ ×4 28	$\overset{1}{6}232$ ×4 928	$\overset{1}{6}232$ ×4 24928

Multiply.

	a	b	c	d	e	f
1.	3 ×2	3000 ×2	2 ×4	2000 ×4	3 ×3	3000 ×3
2.	7 ×3	7000 ×3	8 ×5	8000 ×5	9 ×8	9000 ×8

Multiply.

	a	b	c	d	e
3.	3412 ×2	2018 ×4	1071 ×5	2731 ×3	8021 ×4
4.	1049 ×7	5107 ×8	1614 ×6	1751 ×5	5401 ×9
5.	5671 ×6	5407 ×9	4758 ×7	7034 ×5	6752 ×8

Perfect score: 27   My score: _____

Problem Solving

Solve each problem.

1. Each tank truck can haul 5,000 gallons. How many gallons can be hauled by 9 such trucks?

_____ gallons can be hauled.

2. Sample boxes of soap flakes were given out in 7 cities. In each city 2,500 boxes were given out. How many boxes were given out in all?

_____ boxes were given out.

3. Eight autos are on a freight car. Each auto weighs 1,932 kilograms. What is the weight of all the autos on the freight car?

The combined weight is _____ kilograms.

4. The rail distance between St. Louis and San Francisco is 2,134 miles. How many miles does a train travel on a round trip between those cities?

_____ miles are traveled.

5. An airline attendant made 5 flights last week. The average length of each flight was 1,691 kilometers. How many kilometers did the attendant fly last week?

The attendant flew _____ kilometers.

6. There are 9 machines in a warehouse. Each machine weighs 1,356 kilograms. What is the combined weight of the machines?

The combined weight is _____ kilograms.

7. The Gorman News Agency distributes 6,525 newspapers daily. How many newspapers does it distribute in 6 days?

_____ newspapers are distributed in 6 days.

1.	
2.	**3.**
4.	**5.**
6.	**7.**

Perfect score: 7   My score: _____

62

Lesson 2 Multiplication

Multiply 5372
by 8 ones.

```
  5372
 × 38
 42976
```

Multiply 5372
by 3 tens.

```
   5372
  × 38
  42976
 161160
```

```
   5372
  × 38
  42976  ⎫
 161160  ⎬ Add.
 204136  ⎭
```

Multiply.

	a	b	c	d
1.	4000 ×2	4000 ×20	5000 ×3	5000 ×30
2.	2000 ×40	3000 ×30	7000 ×50	6000 ×90
3.	2031 ×32	3132 ×22	2120 ×34	2314 ×25
4.	4312 ×28	8752 ×19	4321 ×72	3012 ×93
5.	7654 ×81	7542 ×65	8075 ×96	6209 ×58

Perfect score: 20   My score: _____

Problem Solving

Solve each problem.

1. Each day 7,500 tons of ore can be processed. How many tons can be processed in 25 days?

_____ tons can be processed.

2. A large ocean liner can carry 2,047 passengers. How many passengers can it carry on 36 trips?

It can carry _____ passengers on 36 trips.

3. A computer can perform 9,456 computations per second. How many computations can it perform in 1 minute? (1 minute = 60 seconds)

_____ computations can be performed.

4. Mr. LaFong earns $1,399 each month. How much will he earn in 12 months?

He will earn $_____.

5. The Bulls played 82 basketball games last year. The average attendance at each game was 6,547. What was the total attendance?

The total attendance was _____.

6. There are 15 sections of reserved seats. Each section has 1,356 seats. How many reserved seats are there in all?

There are _____ reserved seats.

7. Each day 2,228 cars can be assembled. How many cars can be assembled in 49 days?

_____ cars can be assembled.

8. The Empire State Building is 1,472 feet tall. Mount Everest is 20 times taller than that. How tall is Mount Everest?

Mount Everest is _____ feet tall.

1.	2.
3.	4.
5.	6.
7.	8.

Perfect score: 8   My score: _____

64

Lesson 3   Multiplication

Multiply.

	a	b	c	d
1.	3412 ×2	3127 ×3	8101 ×5	2421 ×4
2.	1307 ×6	8172 ×4	9701 ×9	7061 ×8
3.	1567 ×5	4063 ×7	3579 ×8	8759 ×6
4.	2301 ×23	4753 ×71	4321 ×82	3012 ×93
5.	5324 ×19	2321 ×37	4032 ×28	1202 ×46
6.	5678 ×57	9305 ×78	6078 ×96	8349 ×74

Perfect score: 24   My score: _____

Problem Solving

Solve each problem.

1. Last week an average of 5,112 books a day was checked out of the city library. The library is open 6 days a week. How many books were checked out last week?

_____ books were checked out.

2. Suppose books continue to be checked out at the rate indicated in problem 1. How many books will be checked out in 26 days?

_____ books would be checked out.

3. An auto dealer hopes to sell twice as many cars this year as last year. He sold 1,056 cars last year. How many cars does the dealer hope to sell this year?

The dealer hopes to sell _____ cars.

4. The Humphreys drive an average of 1,245 miles each month. How many miles will they drive in a year? (1 year = 12 months)

They will drive _____ miles.

5. The supermarket sells an average of 1,028 dozen eggs each week. How many dozen eggs will be sold in 6 weeks?

_____ dozen eggs will be sold.

6. The Record Shoppe sells an average of 1,435 records each week. How many records will be sold in 52 weeks?

_____ records will be sold.

7. A certain machine can produce 2,154 items an hour. How many items can be produced in 8 hours?

_____ items can be produced.

1.	
2.	3.
4.	5.
6.	7.

Perfect score: 7   My score: _____

66

Lesson 4   Multiplication

Multiply 512 by 4 ones.	Multiply 512 by 2 tens.	Multiply 512 by 3 hundreds.	
512 ×324 ———— 2048	512 ×324 ———— 2048 10240	512 ×324 ———— 2048 10240 153600	512 ×324 ———— 2048 ⎫ 10240 ⎬ Add. 153600 ⎭ ———— 165888

Multiply.

	a	b	c	d	e
1.	213 ×3	213 ×300	324 ×7	324 ×700	248 ×900
2.	321 ×213	223 ×239	342 ×260	213 ×823	423 ×257
3.	725 ×508	423 ×672	709 ×591	648 ×479	568 ×986

Perfect score: 15   My score: _____

67

Problem Solving

Flight 704 will have 125 passengers.

Solve each problem.

1. Suppose each passenger on Flight 704 has 100 pounds of luggage. What is the total weight of the luggage?

The total weight is _____ pounds.

2. The average weight of each passenger is 135 pounds. What is the total weight of the passengers on Flight 704?

The total weight is _____ pounds.

3. Each passenger on Flight 704 paid $103 for a ticket. How much money was paid in all?

$_____ was paid in all.

4. The air distance from Chicago to Washington, D.C. is 957 kilometers. An airplane will make that flight 365 times this year. How many kilometers will be flown?

_____ kilometers will be flown.

5. A jet was airborne 885 hours last year. Its average speed was 880 kilometers per hour. How far did the jet fly last year?

_____ kilometers were flown last year.

1.

2.

3.

4.

5.

Perfect score: 5   My score: _____

68

Lesson 5   Multiplication

Multiply.

	a	b	c	d
1.	203 ×213	143 ×121	432 ×128	213 ×216
2.	312 ×327	423 ×162	324 ×291	574 ×801
3.	234 ×712	212 ×934	567 ×148	423 ×278
4.	787 ×619	578 ×807	243 ×725	678 ×491
5.	785 ×580	584 ×787	375 ×998	673 ×578

Perfect score: 20   My score: _____

Problem Solving

Solve each problem.

1. Last year a bookstore sold an average of 754 books on each of the 312 days it was open. How many books were sold last year?

_____ books were sold.

2. In one hour, 560 loaves of bread can be baked. How many loaves can be baked in 112 hours?

_____ loaves can be baked.

3. The garment factory can manufacture 960 shirts each day. How many shirts can be manufactured in 260 days?

_____ shirts can be manufactured.

4. Approximately 925 rolls of newsprint are used each week in putting out the daily newspaper. How many rolls of newsprint will be needed in 104 weeks?

_____ rolls of newsprint will be needed.

5. The average American uses 225 liters of water each day. How many liters of water does the average American use in 365 days?

_____ liters of water are used in 365 days.

6. A factory has 109 shipping crates. Each crate can hold 148 pounds of goods. How many pounds of goods can be shipped in all the crates?

_____ pounds of goods can be shipped.

7. A service station sells an average of 965 gallons of gasoline per day. How many gallons will be sold in 365 days?

_____ gallons will be sold.

1.	
2.	3.
4.	5.
6.	7.

Perfect score: 7   My score: _____

70

CHAPTER 6 TEST

Multiply.

	a	b	c	d
1.	2021 ×4	1107 ×6	4321 ×9	6758 ×8
2.	1022 ×23	2121 ×14	3241 ×27	3121 ×38
3.	5264 ×71	2134 ×82	5768 ×67	4938 ×89
4.	324 ×212	243 ×127	321 ×362	212 ×524
5.	584 ×167	778 ×518	432 ×692	789 ×694

Perfect score: 20   My score: _____

PRE-TEST—Temperature and Money

Record the temperature reading shown on each thermometer.

	a	b	c	d

1.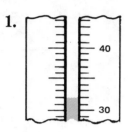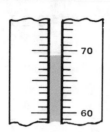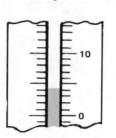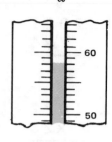

_____° C   _____° F   _____° C   _____° F

Add or subtract.

	a	b	c	d	e
2.	3 7¢ +4 2¢	$ .4 3 +.6 4	1 7¢ 2 6¢ +4 2¢	$2.7 5 6.2 1 +3.6 2	$1 7.3 4 4 5.8 7 +2 5.1 2
3.	7 9¢ −2 7¢	$ .8 3 −.2 4	$4.3 4 −.7 2	$8.6 7 −4.8 9	$2 7.6 4 −1 8.6 9

Multiply.

4.	$.1 6 ×6	$ .4 6 ×7	$2.4 3 ×2	$1.1 7 ×1 9	$1 7.4 5 ×2 5

Solve each problem.

5. A suit costs $99.88. A sports coat costs $49.95. How much more does the suit cost than the sports coat?

| **5.** | **6.** |

The suit costs $_____ more.

6. You spend $1.98 in one store and $.79 in another. How much do you spend in both stores?

You spend $_____ in both stores.

Perfect score: 21   My score: _____

Lesson 1   Temperature

Celsius

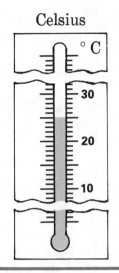

Fahrenheit

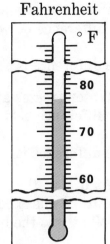

Both thermometers show the same temperature.

The temperature on the Celsius thermometer is 25 degrees. This can be written 25° C.

The temperature on the Fahrenheit thermometer is 77 degrees. This can be written as 77° F.

Record the temperature reading shown on each thermometer.

	a	*b*	*c*	*d*

1.

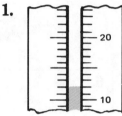

_____ ° C

_____ ° C

_____ ° C

_____ ° C

2.

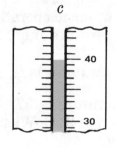
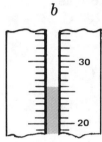
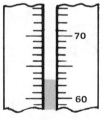
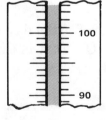

_____ ° F

_____ ° F

_____ ° F

_____ ° F

3.

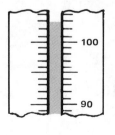
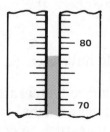

_____ ° C

_____ ° F

_____ ° C

_____ ° F

Perfect score: 12   My score: _____

73

Problem Solving

Solve each problem.

1. Water freezes at 0° C and boils at 100° C. What is the difference between those two temperatures?

The difference is _____ degrees.

2. Water freezes at 32° F and boils at 212° F. What is the difference between those two temperatures?

The difference is _____ degrees.

3. At 6 A.M. the temperature was 56° F. The high temperature was expected to be 35 degrees warmer than that. What was the high temperature expected to be?

_____ ° F was the expected high temperature.

4. On June 1 the water temperature at the lake was 10° C. During June the water temperature rose 10° C. What was the water temperature July 1?

The temperature of the water was _____ ° C.

5. Normal body temperature is 37° C. When Rita was ill, her temperature rose 2° C. What was her temperature when she was ill?

Her temperature was _____ ° C when she was ill.

6. The high temperature for the day was 25° C. The low temperature was 16° C. What is the difference between those two temperatures?

The difference is _____ degrees.

7. During a windy day the windchill was 6° F. With no wind, the temperature would have been 29 degrees warmer. What would have been the temperature with no wind?

The temperature would have been _____ ° F.

8. Yesterday Adam had a temperature of 104° F. Today his temperature is near normal at 99° F. By how many degrees did his temperature go down?

His temperature went down _____ degrees.

1.	2.
3.	4.
5.	6.
7.	8.

Perfect score: 8   My score: _____

74

Lesson 2   Money

NAME _____

1 cent = 1¢ or $.01
3 cents = ___3¢___ or ___$.03___
65 cents = ___65¢___ or ___$.65___

6 cents = _____ ¢ or $_____

98 cents = _____ ¢ or $_____

1 dollar = $1.00
3 dollars and 2 cents = ___$3.02___
4 dollars and 59 cents = ___$4.59___

6 dollars and 3 cents = $_____

5 dollars and 72 cents = $_____

Complete the following.

	a	b	c
1.	5 cents = _____ ¢	25¢ = $_____	$.83 = _____ ¢
2.	10 cents = $_____	50¢ = $_____	$.04 = _____ ¢
3.	25 cents = _____ ¢	75¢ = $_____	$.29 = _____ ¢
4.	50 cents = $_____	10¢ = $_____	$.06 = _____ ¢
5.	85 cents = _____ ¢	95¢ = $_____	$.60 = _____ ¢
6.	100 cents = $_____	5¢ = $_____	$.99 = _____ ¢

Complete the following.

	a	b
7.	4 dollars and 8 cents = $_____	$6.25 = 6 dollars and _____ cents
8.	7 dollars and 63 cents = $_____	$3.75 = _____ dollars and 75 cents
9.	3 dollars and 9 cents = $_____	$7.05 = 7 dollars and _____ cents
10.	6 dollars and 19 cents = $_____	$9.65 = _____ dollars and 65 cents
11.	5 dollars and 79 cents = $_____	$4.19 = _____ dollars and _____ cents
12.	18 dollars and 75 cents = $_____	$8.69 = _____ dollars and _____ cents

Perfect score: 32   My score: _____

Lesson 3   Money

NAME _____

Add the numbers.

```
  25¢      $ .85
  45¢       2.08
+ 19¢     + 3.76
─────     ───────
  89¢      $6.69
```

Write ¢ or $ and a decimal
point in the answer.

Subtract the numbers.

```
  72¢      $12.07
− 26¢     − 4.83
─────     ───────
  46¢      $ 7.24
```

Write ¢ or $ and a decimal
point in the answer.

Add or subtract.

	a	b	c	d	e
1.	2 3¢ + 4 4¢	4 7¢ + 2 5¢	$.4 6 + .7 3	$5.4 7 + 8.2 1	$3 6.9 5 + 7 2.0 2
2.	7 9¢ − 2 3¢	5 6¢ − 2 7¢	$1.2 7 − .5 3	$4.6 7 − 2.8 9	$3 6.7 8 − 7.9 9
3.	1 4¢ + 7 1¢	6¢ + 8 7¢	$.5 7 + .6 8	$5.2 5 + 9.4 6	$1 6.9 6 + 2 7.4 5
4.	8 8¢ − 6 9¢	9 2¢ − 8 9¢	$2.6 4 − .5 7	$6.2 7 − 2.8 9	$4 9.7 8 − 1 8.8 9
5.	1 2¢ 3 9¢ + 2 4¢	4 3¢ 2 7¢ + 2 6¢	$.7 5 .6 5 + .9 7	$ .1 2 4.6 9 + 5.8 7	$4 7.5 2 8 9.2 5 + 6 7.4 7
6.	$2.4 6 − .8 7	$1.5 7 − .9 9	$3.0 7 − 1.8 5	$7.0 0 − 2.4 8	$6 0.4 7 − 2 7.5 9

Perfect score: 30   My score: _____

Lesson 4   Money

$.19
×4
——
$.76

Multiply the numbers.
Write $ and a decimal
point in the answer.

Make sure there are
two digits to the
right of the decimal
point.

$3.24
×46
——
1944
12960
——
$149.04

Multiply.

	a	b	c	d	e
1.	$.2 1 × 4	$.1 3 × 5	$.2 6 ×3	$1.7 2 ×4	$1 6.2 3 ×3
2.	$.3 1 × 6	$.5 2 × 7	$.5 8 ×6	$2.4 7 ×4	$2 5.7 9 ×3
3.	$.3 2 × 1 2	$.4 3 × 2 3	$.3 4 ×5 2	$4.4 9 ×3 6	$4 3.7 5 ×2 4
4.	$2.3 8 ×4 2 6	$3.0 9 ×3 2 9	$1.2 4 ×1 0 2	$4.2 9 ×3 1 7	$7 5.9 6 ×4 4

Perfect score: 20   My score: _____

Problem Solving

Solve each problem.

1. Find the cost of these four items: soap, $.58; soup, $.55; napkins, $.87; and toothpicks, $.39.

The total cost is $_____.

2. A quart of motor oil sells for $1.49 at a service station. It costs $.99 at a discount store. What is the difference between those prices?

The difference is _____¢.

3. Spark plugs are on sale at $2.19 each. What is the cost of 8 spark plugs?

The cost is $_____.

4. It costs $19.75 to go to River City by train. It costs $16.95 to go by bus. How much cheaper is it to go by bus?

It is $_____ cheaper.

5. Tires cost $89.89 each. What is the cost of 4 tires?

The cost is $_____.

6. A ticket for a bleacher seat costs $3.75. A ticket for a box seat costs $9.50. How much more does it cost to sit in a box seat than in a bleacher seat?

It costs $_____ more.

7. How much money have the four pupils named in the table saved?

They have saved $_____.

8. How much more money has Cary saved than Jerri?

Cary has saved $_____ more.

Pupil	Amount saved
Barry	$27.49
Cary	$33.14
Jerri	$29.36
Mary	$28.76

1.	2.
3.	4.
5.	6.
7.	8.

Perfect score: 8   My score: _____

78

CHAPTER 7 TEST

Record the temperature reading shown on each thermometer.

a	b	c	d

1.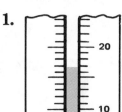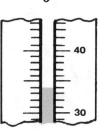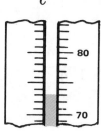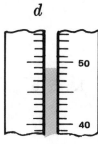

_____ ° C   _____ ° F   _____ ° C   _____ ° F

Add.

a	b	c	d	e

2.
```
    7 2¢        5 6¢        $.2 9       $9.7 9      $2 4.5 9
  +1 7¢       +2 5¢       +.3 5       +4.8 5        1 9.5 7
                                                  +2 8.3 6
```

Subtract.

3.
```
    8 7¢        4 8¢        $.7 1       $8.5 2      $1 0.0 0
  -5 5¢       -3 9¢       -.5 4       -5.4 9       - 3.3 7
```

Multiply.

4.
```
   $.2 9       $.4 6       $1.3 9      $4.6 9      $3.2 4
   × 3         × 6         × 8         ×1 2        ×2 4
```

Solve the problem.

5. Mr. Kingman bought a bat for $24.49, a ball for $7.98, and a glove for $37.56. How much did he spend in all?

5.

Mr. Kingman spent $_____ in all.

Perfect score: 20   My score: _____

PRE-TEST—Division

Divide.

	a	*b*	*c*	*d*	*e*
1.	5⟌5	1⟌4	3⟌1 2	2⟌4	4⟌2 8
2.	1⟌0	2⟌1 8	4⟌8	3⟌2 4	5⟌3 5
3.	6⟌2 4	6⟌1 2	6⟌4 8	6⟌6	6⟌3 0
4.	6⟌4 2	6⟌1 8	6⟌3 6	6⟌5 4	6⟌0
5.	7⟌5 6	7⟌4 2	7⟌0	7⟌2 8	7⟌1 4
6.	7⟌6 3	7⟌2 1	7⟌4 9	7⟌7	7⟌3 5
7.	8⟌5 6	8⟌4 0	8⟌2 4	8⟌7 2	8⟌8
8.	8⟌1 6	8⟌6 4	8⟌0	8⟌3 2	8⟌4 8
9.	9⟌5 4	9⟌9	9⟌7 2	9⟌3 6	9⟌1 8
10.	9⟌8 1	9⟌6 3	9⟌2 7	9⟌0	9⟌4 5

Perfect score: 50   My score: _____

Lesson 1 Division

```
  ┌─► 4
  │ 3⟌12 ─► Find the 12 in
  └─ ─► the  3  -column.
  └ ─ ─ ─ ─ The quotient is named
            in the ▨ at the end
            of this row.
```

Use the table to divide.

5⟌30

×	0	1	2	3	4	5	6	7	8	9
0	0	0	0	0	0	0	0	0	0	0
1	0	1	2	3	4	5	6	7	8	9
2	0	2	4	6	8	10	12	14	16	18
3	0	3	6	9	12	15	18	21	24	27
④	0	4	8	12	16	20	24	28	32	36
5	0	5	10	15	20	25	30	35	40	45
⑥	0	6	12	18	24	30	36	42	48	54
7	0	7	14	21	28	35	42	49	56	63
8	0	8	16	24	32	40	48	56	64	72
9	0	9	18	27	36	45	54	63	72	81

Divide.

	a	b	c	d	e	f
1.	3⟌15	4⟌12	1⟌7	5⟌25	2⟌12	4⟌28
2.	5⟌35	2⟌14	3⟌18	4⟌20	1⟌9	3⟌9
3.	4⟌32	5⟌20	1⟌6	3⟌12	2⟌10	3⟌21
4.	2⟌16	3⟌3	4⟌16	1⟌8	4⟌0	5⟌10
5.	3⟌24	5⟌0	2⟌8	4⟌36	5⟌15	3⟌27
6.	4⟌24	2⟌18	5⟌40	3⟌0	1⟌5	5⟌45
7.	1⟌4	2⟌6	5⟌30	5⟌5	4⟌8	3⟌6

Perfect score: 42   My score: _____

81

Problem Solving

Solve each problem.

1. There are 24 cars in the parking lot. There are 3 rows of cars. Each row has the same number of cars. How many cars are in each row?

There are _____ cars.

There are _____ rows.

There are _____ cars in each row.

2. Five people went to the zoo. The tickets cost $25 in all. Each ticket cost the same amount. How much did each ticket cost?

Each ticket cost $_____.

3. There are 16 people playing ball. There are 2 teams. Each team has the same number of players. How many people are on each team?

_____ people are on each team.

4. Four people are seated at each table. There are 16 people in all. How many tables are there?

There are _____ tables.

5. Two people are in each boat. There are 16 people in all. How many boats are there?

There are _____ boats.

1.

2.

3.

4.

5.

Perfect score: 7   My score: _____

Lesson 2   Division

$$5 \dashrightarrow 5$$
$$\times 6 \dashrightarrow 6\overline{)30}$$
$$\overline{30} \dashrightarrow$$

$$5 \dashrightarrow 5$$
$$\times 7 \dashrightarrow 7\overline{)35}$$
$$\overline{35} \dashrightarrow$$

$6 \times 5 = 30$, so $30 \div 6 = \underline{\quad 5 \quad}$.

$7 \times 5 = 35$, so $35 \div 7 = \underline{\qquad}$.

Complete the following.

	a	b	c	d
1.	$\begin{array}{r} 6 \\ \times 6 \\ \hline 36 \end{array}$ so $6\overline{)36.}$	$\begin{array}{r} 7 \\ \times 6 \\ \hline 42 \end{array}$ so $6\overline{)42.}$	$\begin{array}{r} 8 \\ \times 6 \\ \hline 48 \end{array}$ so $6\overline{)48.}$	$\begin{array}{r} 9 \\ \times 6 \\ \hline 54 \end{array}$ so $6\overline{)54.}$
2.	$\begin{array}{r} 6 \\ \times 7 \\ \hline 42 \end{array}$ so $7\overline{)42.}$	$\begin{array}{r} 7 \\ \times 7 \\ \hline 49 \end{array}$ so $7\overline{)49.}$	$\begin{array}{r} 8 \\ \times 7 \\ \hline 56 \end{array}$ so $7\overline{)56.}$	$\begin{array}{r} 9 \\ \times 7 \\ \hline 63 \end{array}$ so $7\overline{)63.}$

Divide.

	a	b	c	d	e
3.	$6\overline{)54}$	$7\overline{)14}$	$6\overline{)30}$	$7\overline{)35}$	$6\overline{)6}$
4.	$7\overline{)49}$	$6\overline{)42}$	$7\overline{)42}$	$6\overline{)0}$	$7\overline{)21}$
5.	$6\overline{)18}$	$7\overline{)7}$	$6\overline{)36}$	$7\overline{)56}$	$6\overline{)24}$
6.	$7\overline{)28}$	$6\overline{)48}$	$7\overline{)0}$	$6\overline{)12}$	$7\overline{)63}$
7.	$4\overline{)16}$	$4\overline{)28}$	$5\overline{)20}$	$5\overline{)35}$	$4\overline{)36}$

Perfect score: 33   My score: _____

Problem Solving

Solve each problem.

1. Mrs. Shields had 24 plants. She put them into rows of 6 plants each. How many rows were there?

Mrs. Shields had _____ plants.

She put them into rows of _____ plants each.

She had _____ rows of plants.

2. 28 pupils sat at 7 tables. The same number of pupils sat at each table. How many pupils sat at each table?

There were _____ pupils.

There were _____ tables.

_____ pupils sat at each table.

3. Pencils cost 7¢ each. Marcella has 63¢. How many pencils can she buy?

She can buy _____ pencils.

4. Mr. Rojas put 56 books into stacks of 7 books each. How many stacks of books did he have?

He had _____ stacks of books.

5. Marion Street has 54 streetlights. Each block has 6 lights. How many blocks are there?

There are _____ blocks.

6. There are 35 days before Ed's birthday. How many weeks is it before his birthday? (7 days = 1 week)

There are _____ weeks before Ed's birthday.

7. There are 48 items to be packed. Six items can be packed in each box. How many boxes are needed?

_____ boxes are needed.

1.	
2.	
3.	**4.**
5.	**6.**
7.	

Perfect score: 11   My score: _____

84

Lesson 3 Division

```
4 --------→  4
×8 -----→ 8|32
32 --------⌐
```

```
5 --------→  5
×9 -----→ 9|45
45 --------⌐
```

$8 \times 4 = 32$, so $32 \div 8 =$ ___4___.

$9 \times 5 = 45$, so $45 \div 9 =$ _____.

Complete the following.

	a	b	c	d

1.

```
    6                7                8                9
   ×8   so  8|48.   ×8   so  8|56.   ×8   so  8|64.   ×8   so  8|72.
   48               56               64               72
```

2.

```
    6                7                8                9
   ×9   so  9|54.   ×9   so  9|63.   ×9   so  9|72.   ×9   so  9|81.
   54               63               72               81
```

Divide.

	a	b	c	d	e

3. 8|8   9|1 8   8|5 6   9|2 7   8|3 2

4. 9|4 5   8|6 4   9|0   8|2 4   9|5 4

5. 8|7 2   9|3 6   8|1 6   9|9   8|4 0

6. 9|6 3   8|0   9|7 2   8|4 8   9|8 1

7. 7|4 2   6|4 8   7|5 6   6|5 4   7|6 3

Perfect score: 33   My score: _____

85

Problem Solving

Solve each problem.

1. A checkerboard has 64 squares. Each of the 8 rows has the same number of squares. How many squares are in each row?

_____ squares are in each row.

2. 36 children came to the park to play ball. How many teams of 9 players each could be formed?

_____ teams could be formed.

3. Each washer load weighs 9 pounds. How many washer loads are there in 54 pounds of laundry?

_____ loads are in 54 pounds of laundry.

4. There are 48 apartments on 8 floors. There is the same number of apartments on each floor. How many apartments are there on each floor?

_____ apartments are on each floor.

5. 9 books weigh 18 pounds. Each book has the same weight. How much does each book weigh?

Each book weighs _____ pounds.

6. Miss McKee can type 8 pages an hour. How long would it take her to type 24 pages?

It would take her _____ hours.

7. How many tosses can you make for 72¢?

_____ tosses can be made.

Bean Bag Toss
8¢ per toss

6
7 8

8. Mabelle tossed 7 bags through the same hole. Her score was 49. How many points did she score on each toss?

She scored _____ points on each toss.

1.	2.
3.	4.
5.	6.
7.	8.

Perfect score: 8   My score: _____

86

Lesson 4   Division

Complete the following.

	a	b	c	d

1.
$$\begin{array}{r} 4 \\ \times 9 \\ \hline 36 \end{array}$$ so $9\overline{)36}.$
$$\begin{array}{r} 7 \\ \times 2 \\ \hline 14 \end{array}$$ so $2\overline{)14}.$
$$\begin{array}{r} 8 \\ \times 1 \\ \hline 8 \end{array}$$ so $1\overline{)8}.$
$$\begin{array}{r} 7 \\ \times 6 \\ \hline 42 \end{array}$$ so $6\overline{)42}.$

2.
$$\begin{array}{r} 6 \\ \times 7 \\ \hline 42 \end{array}$$ so $7\overline{)42}.$
$$\begin{array}{r} 8 \\ \times 3 \\ \hline 24 \end{array}$$ so $3\overline{)24}.$
$$\begin{array}{r} 6 \\ \times 4 \\ \hline 24 \end{array}$$ so $4\overline{)24}.$
$$\begin{array}{r} 0 \\ \times 5 \\ \hline 0 \end{array}$$ so $5\overline{)0}.$

Divide.

a	b	c	d	e
3. $4\overline{)36}$	$9\overline{)72}$	$6\overline{)54}$	$5\overline{)10}$	$8\overline{)56}$
4. $2\overline{)6}$	$4\overline{)20}$	$7\overline{)28}$	$1\overline{)4}$	$6\overline{)30}$
5. $7\overline{)56}$	$6\overline{)18}$	$3\overline{)0}$	$9\overline{)54}$	$8\overline{)40}$
6. $1\overline{)6}$	$9\overline{)18}$	$5\overline{)20}$	$7\overline{)14}$	$4\overline{)12}$
7. $2\overline{)2}$	$2\overline{)18}$	$8\overline{)24}$	$6\overline{)6}$	$1\overline{)2}$
8. $7\overline{)0}$	$3\overline{)12}$	$3\overline{)6}$	$9\overline{)0}$	$5\overline{)30}$
9. $5\overline{)40}$	$8\overline{)8}$	$2\overline{)10}$	$1\overline{)0}$	$4\overline{)4}$
10. $3\overline{)18}$	$7\overline{)35}$	$8\overline{)32}$	$9\overline{)27}$	$2\overline{)16}$

Perfect score: 48   My score: _____

Problem Solving

Solve each problem.

1. There are 45 school days left before vacation. There are 5 school days each week. How many weeks are left before vacation?

There are _____ weeks left.

2. There are 35 seats in Mrs. Champney's room. The seats are arranged in 7 rows with the same number in each row. How many seats are there in each row?

There are _____ seats in each row.

3. Diane bought 21 feet of material. How many yards of material did she buy? (3 feet = 1 yard)

Diane bought _____ yards of material.

4. Forty-eight cars are parked in a parking lot. The cars are parked in 6 rows with the same number in each row. How many cars are parked in each row?

_____ cars are parked in each row.

5. There are 36 seats on a bus and 4 seats per row. How many rows of seats are there on the bus?

There are _____ rows of seats on the bus.

6. Mr. Woods works a 6-day week. He works 54 hours each week and the same number of hours each day. How many hours does he work each day?

He works _____ hours each day.

7. It takes 9 minutes to assemble a certain item. How many items can be assembled in 54 minutes?

_____ items can be assembled.

8. There are 32 girls in a relay race. Four run on each team. How many teams are there?

There are _____ teams.

1.	2.
3.	4.
5.	6.
7.	8.

Perfect score: 8   My score: _____

88

CHAPTER 8 TEST

Divide.

	a	*b*	*c*	*d*	*e*
1.	8$\overline{)16}$	5$\overline{)15}$	2$\overline{)0}$	6$\overline{)48}$	9$\overline{)81}$
2.	4$\overline{)8}$	1$\overline{)1}$	3$\overline{)3}$	8$\overline{)72}$	7$\overline{)63}$
3.	3$\overline{)18}$	4$\overline{)32}$	7$\overline{)42}$	2$\overline{)4}$	5$\overline{)5}$
4.	7$\overline{)49}$	5$\overline{)25}$	6$\overline{)36}$	9$\overline{)9}$	2$\overline{)8}$
5.	6$\overline{)42}$	2$\overline{)12}$	8$\overline{)0}$	1$\overline{)3}$	4$\overline{)28}$
6.	3$\overline{)15}$	6$\overline{)24}$	1$\overline{)5}$	8$\overline{)32}$	4$\overline{)24}$
7.	5$\overline{)35}$	1$\overline{)7}$	6$\overline{)12}$	7$\overline{)7}$	9$\overline{)27}$
8.	3$\overline{)9}$	4$\overline{)16}$	2$\overline{)16}$	3$\overline{)21}$	1$\overline{)9}$
9.	8$\overline{)48}$	7$\overline{)21}$	5$\overline{)45}$	9$\overline{)45}$	6$\overline{)0}$
10.	9$\overline{)63}$	4$\overline{)0}$	3$\overline{)27}$	7$\overline{)35}$	8$\overline{)64}$

Perfect score: 50   My score: _____

PRE-TEST—Division

Divide.

	a	*b*	*c*	*d*	*e*
1.	3)‾2‾6‾	5)‾4‾7‾	6)‾4‾9‾	8)‾6‾2‾	9)‾7‾7‾
2.	2)‾2‾8‾	4)‾4‾8‾	3)‾9‾3‾	7)‾8‾4‾	6)‾9‾6‾
3.	5)‾6‾7‾	8)‾9‾3‾	9)‾9‾7‾	6)‾8‾7‾	7)‾9‾7‾
4.	3)‾1‾8‾6‾	4)‾2‾3‾6‾	7)‾1‾6‾1‾	9)‾4‾2‾5‾	8)‾6‾1‾2‾
5.	3)‾3‾6‾9‾	4)‾8‾4‾0‾	2)‾9‾6‾4‾	5)‾6‾9‾6‾	7)‾8‾9‾8‾

Perfect score: 25   My score: _____

90

Lesson 1   Division

Study how to divide 32 by 6.

×	1	2	3	4	5	6
6	6	12	18	24	30	36

32 is between 30 and 36, so 32 ÷ 6 is between 5 and 6. The ones digit is 5.

Since 32 − 30 = 2 and 2 is less than 6, the **remainder** 2 is recorded like this:

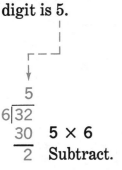

```
   5
6) 32
   30    5 × 6
    2    Subtract.
```

```
   5 r2
6) 32  ↑
   30  |
    2 -┘
```

Study how to divide 43 by 9.

×	1	2	3	4	5	6
9	9	18	27	36	45	54

43 is between 36 and 45, so 43 ÷ 9 is between 4 and 5. The ones digit is 4.

Since 43 − 36 = 7 and 7 is less than 9, the **remainder** is 7.

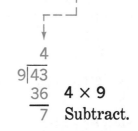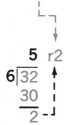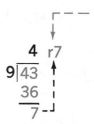

```
   4
9) 43
   36    4 × 9
    7    Subtract.
```

```
   4 r7
9) 43  ↑
   36  |
    7 -┘
```

Divide.

	a	*b*	*c*	*d*	*e*
1.	5⟌27	8⟌47	2⟌17	9⟌46	6⟌38
2.	7⟌61	9⟌67	5⟌49	3⟌23	8⟌78
3.	4⟌38	6⟌45	8⟌63	2⟌19	7⟌38

Perfect score: 15   My score: _____

Problem Solving

Solve each problem.

1. How many pieces of rope 4 meters long can be cut from 15 meters of rope? How much rope will be left over?

_____ pieces can be cut.

_____ meters will be left over.

2. Dale has 51¢. Magic rings cost 8¢ each. How many magic rings can he buy? How much money will he have left?

Dale can buy _____ magic rings.

He will have _____ ¢ left.

3. Maria has a board that is 68 inches long. How many pieces 9 inches long can be cut from this board? How long is the piece that is left?

_____ pieces can be cut.

_____ inches will be left.

4. 28 liters of water was used to fill buckets that hold 5 liters each. How many buckets were filled? How much water was left over?

_____ buckets were filled.

_____ liters of water was left over.

5. It takes 8 minutes to make a doodad. How many doodads can be made in 60 minutes? How much time would be left to begin making another doodad?

_____ doodads can be made.

_____ minutes would be left.

1.

2.

3.

4.

5.

Perfect score: 10   My score: _____

92

Lesson 2   Division

Study how to divide 72 by 3.

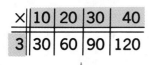

×	10	20	30	40
3	30	60	90	120

72 is between 60 and 90, so
$72 \div 3$ is between 20 and 30.
The tens digit is 2.

$$\begin{array}{r} 2 \\ 3\overline{)72} \\ 60 \quad 20 \times 3 \\ \hline 12 \quad \text{Subtract.} \end{array}$$

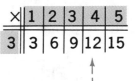

×	1	2	3	4	5
3	3	6	9	12	15

$4 \times 3 = 12$, so the ones
digit is 4.

$$\begin{array}{r} 24 \\ 3\overline{)72} \\ 60 \\ \hline 12 \\ 12 \quad 4 \times 3 \\ \hline 0 \quad \text{Subtract.} \end{array}$$

Divide.

	a	*b*	*c*	*d*	*e*
1.	2)26	4)48	5)55	3)96	4)88
2.	4)47	6)69	3)65	7)96	8)99
3.	7)91	9)96	3)87	8)97	4)92

Perfect score: 15   My score: _____

Problem Solving

Solve each problem.

1. 99 pupils are to be separated into 9 groups. The same number of pupils is to be in each group. How many pupils will be in each group?

_____ pupils will be in each group.

2. There are 94 grapefruit in a crate. How many bags of 6 grapefruit each can be filled by using the grapefruit from 1 crate? How many grapefruit will be left over?

_____ bags can be filled.

_____ grapefruit will be left over.

3. Ms. McClean's lot is 60 feet wide. What is the width of the lot in yards? (3 feet = 1 yard)

Her lot is _____ yards wide.

4. There are 96 fluid ounces of punch in a bowl. How many 7-ounce glasses can be filled? How many fluid ounces will be left over?

_____ glasses can be filled.

_____ fluid ounces will be left over.

5. Toy cars are packed into boxes of 8 cars each. How many boxes will be needed to pack 96 toy cars?

_____ boxes will be needed.

6. You have 64 pennies to exchange for nickels. How many nickels will you get? How many pennies will be left?

You will get _____ nickels.

_____ pennies will be left.

1.	2.
3.	4.
5.	6.

Perfect score: 9   My score: _____

94

Lesson 3 Division

Divide.

	a	*b*	*c*	*d*	*e*
1.	3⟌2 3	5⟌4 6	7⟌6 7	9⟌8 5	8⟌6 9
2.	2⟌2 8	3⟌3 6	4⟌4 4	3⟌9 3	4⟌8 4
3.	5⟌8 5	6⟌9 6	8⟌9 6	7⟌9 8	9⟌9 0
4.	7⟌7 4	5⟌5 7	8⟌9 9	2⟌8 5	3⟌6 5
5.	3⟌7 9	5⟌6 7	6⟌9 7	4⟌9 4	7⟌8 9

Perfect score: 25   My score: _____

Problem Solving

Solve each problem.

1. Paul has 75 stamps. His album will hold 9 stamps per page. How many pages can he fill? How many stamps will be left over?

He can fill _____ pages.

_____ stamps will be left over.

2. Helena has 96 stamps. Her album will hold 6 stamps per page. How many pages can she fill?

She can fill _____ pages.

3. 77 stamps are given to a stamp club. Each of the 7 members is to receive the same number of stamps. How many stamps will each member get?

Each of the members will get _____ stamps.

4. Anne has 87 cents. What is the greatest number of nickels she could have? What is the least number of pennies she could have?

She could have at most _____ nickels.

She could have as few as _____ pennies.

5. 54 pupils are to be separated into teams of 9 pupils each. How many teams can be formed?

_____ teams can be formed.

6. Rex has 63 empty bottles. He wants to put them into cartons of 8 bottles each. How many cartons can he fill? How many bottles will be left over?

_____ cartons can be filled.

_____ bottles will be left over.

1.	2.
3.	**4.**
5.	**6.**

Perfect score: 9   My score: _____

96

Lesson 4   Division

Study how to divide 263 by 5.

×	10	20	30	40	50	60
5	50	100	150	200	250	300

×	1	2	3	4	5
5	5	10	15	20	25

Since $100 \times 5 = 500$ and 500 is greater than 263, there is no hundreds digit.

$$5\overline{)263}$$

263 is between 250 and 300.
$263 \div 5$ is between 50 and 60.
The tens digit is 5.

```
      5
 5)263
   250    50 × 5
    13    Subtract.
```

13 is between 10 and 15.
$13 \div 5$ is between 2 and 3.
The ones digit is 2.

```
     52 r3
 5)263
   250
    13
    10    2 × 5
     3    Subtract.
```

Divide.

	a	b	c	d	e
1.	$4\overline{)248}$	$6\overline{)366}$	$3\overline{)189}$	$7\overline{)266}$	$8\overline{)472}$
2.	$9\overline{)547}$	$2\overline{)121}$	$5\overline{)308}$	$6\overline{)374}$	$4\overline{)341}$
3.	$8\overline{)735}$	$3\overline{)252}$	$9\overline{)479}$	$7\overline{)378}$	$5\overline{)473}$

Perfect score: 15   My score: _____

Problem Solving

Solve each problem.

1. A truck driver drove from Chicago to St. Louis in 6 hours. The same distance was driven each hour. How many kilometers were driven each hour?

_____ kilometers were driven each hour.

2. A bus left New York City and arrived in Baltimore 4 hours later. The bus went the same distance each hour. How many kilometers were traveled each hour?

_____ kilometers were traveled each hour.

3. Suppose the same distance is driven each hour. How many kilometers must be driven each hour to go from Los Angeles to San Francisco in 8 hours?

_____ kilometers must be driven each hour.

4. The Jetts plan to drive from Los Angeles to Phoenix in 7 hours. Suppose they travel the same distance each hour. How many kilometers must they travel each hour?

_____ kilometers must be driven each hour.

5. Suppose the same distance is traveled each hour. How many kilometers must be traveled each hour to go from Chicago to Pittsburgh in 9 hours?

_____ kilometers must be traveled each hour.

1.	
2.	**3.**
4.	**5.**

Perfect score: 5   My score: _____

Lesson 5   Division

Study how to divide 813 by 4.

×	100	200	300
4	400	800	1200

813 is between 800 and 1200, so 813 ÷ 4 is between 200 and 300. The hundreds digit is 2.

```
  2
4│813
  800    200 × 4
  ───
   13    Subtract.
```

Since 10 × 4 = 40 and 40 is greater than 13, the tens digit is 0.

```
  20
4│813
  800
  ───
   13
    0   0 × 4
  ───
   13   Subtract.
```

×	1	2	3	4
4	4	8	12	16

13 is between 12 and 16, so 13 ÷ 4 is between 3 and 4. The ones digit is 3.

```
  203 r1
4│813
  800
  ───
   13
    0
  ───
   13
   12   3 × 4
  ───
    1   Subtract.
```

Divide.

	a	b	c	d	e
1.	2│4 6 8	4│4 7 2	3│6 0 9	5│5 8 5	7│8 8 2
2.	8│8 7 6	6│7 9 4	9│9 7 9	2│9 8 7	5│5 9 3
3.	6│8 4 2	3│9 4 9	7│8 7 5	4│8 7 9	8│9 9 2

Perfect score: 15   My score: _____

Problem Solving

Solve each problem.

1. Mrs. Steel needs 960 trading stamps to fill a book. These stamps will fill 8 pages with the same number of stamps on each page. How many stamps are needed to fill each page?

_____ stamps are needed to fill each page.

2. There are 576 pencils in 4 cases. There are the same number of pencils in each case. How many pencils are in each case?

There are _____ pencils in each case.

3. There are 532 apples in all. How many sacks of 5 apples each can be filled? How many apples will be left over?

_____ sacks can be filled.

_____ apples will be left over.

4. At a factory, 968 items were manufactured during an 8-hour shift. The same number was manufactured each hour. How many items were manufactured each hour?

_____ items were manufactured each hour.

5. The Chesapeake Bridge is 540 feet long. It consists of 4 sections of the same length. How long is each section?

Each section is _____ feet long.

6. A full load for a dry cleaning machine is 5 suits. There are 624 suits to be cleaned. How many full loads will there be? How many suits will be in the partial load?

There will be _____ full loads.

There will be _____ suits in the partial load.

7. A carpenter uses 5 nails to shingle one square foot of roof. At that rate, how many square feet of roof could be shingled by using 750 nails?

_____ square feet could be shingled.

1.	
2.	**3.**
4.	**5.**
6.	**7.**

Perfect score: 9   My score: _____

100

Lesson 6   Division

Divide.

	a	*b*	*c*	*d*	*e*
1.	2⟌126	6⟌486	3⟌249	7⟌553	8⟌624
2.	5⟌473	4⟌357	9⟌758	6⟌525	3⟌269
3.	3⟌693	2⟌816	4⟌856	5⟌925	7⟌791
4.	9⟌969	6⟌797	8⟌953	7⟌899	5⟌869

Perfect score: 20   My score: _____

Problem Solving

Solve each problem.

1. A rocket used 945 kilograms of fuel in 9 seconds during lift-off. The same amount of fuel was used each second. How many kilograms of fuel were used each second?

_____ kilograms were used each second.

2. A team for a relay race consists of 4 members who run the same distance. How far would each member run in an 880-yard relay?

Each team member would run _____ yards.

3. There are 435 folding chairs in all. How many rows of 9 chairs each can be formed? How many chairs will be left over?

_____ rows can be formed.

_____ chairs will be left over.

4. 8 boxes of freight weigh 920 kilograms. Each box holds the same amount of freight. How much does each box weigh?

Each box weighs _____ kilograms.

5. A person's weight on Earth is 6 times more than on the moon. How much would a person who weighs 180 pounds on Earth weigh on the moon?

The person would weigh _____ pounds.

6. There are 627 items to be packed. The items are to be packed 6 to a box. How many boxes can be filled? How many items will be left over?

_____ boxes will be filled.

_____ items will be left over.

7. The weight of four players on a team is 292 kilograms. Suppose each player weighs the same amount. How much would each player weigh?

Each player would weigh _____ kilograms.

1.	
2.	**3.**
4.	**5.**
6.	**7.**

Perfect score: 9   My score: _____

102

CHAPTER 9 TEST

Divide.

	a	b	c	d	e
1.	4)26	6)39	7)68	8)88	2)86
2.	5)90	6)78	3)68	4)87	8)98
3.	2)104	6)246	3)219	5)285	8)672
4.	4)363	3)278	5)427	9)627	8)465
5.	9)981	4)892	7)952	4)843	8)986

Perfect score: 25   My score: _____

PRE-TEST—Division

Divide.

	a	*b*	*c*	*d*
1.	3)249	5)452	8)4893	6)4257
2.	4)8408	9)9081	7)9147	5)6724
3.	2)1268	4)3282	6)2416	9)1764
4.	5)5650	3)3692	7)8435	8)9472

Perfect score: 16   My score: _____

Lesson 1   Division

Think about each digit in the quotient.

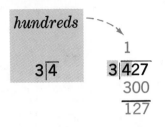
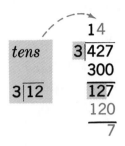
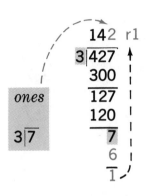

Divide.

	a	b	c	d	e

1. $2\overline{)268}$ $3\overline{)459}$ $5\overline{)705}$ $6\overline{)965}$ $2\overline{)483}$

2. $6\overline{)555}$ $9\overline{)421}$ $7\overline{)686}$ $8\overline{)406}$ $3\overline{)254}$

3. $7\overline{)476}$ $3\overline{)654}$ $5\overline{)982}$ $9\overline{)734}$ $4\overline{)502}$

Perfect score: 15   My score: _____

Lesson 2 Division

Think about each digit in the quotient.

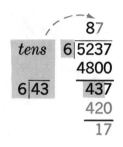

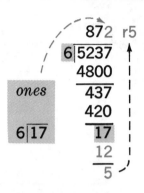

Divide.

	a	b	c	d	e
1.	2)1248	7)4291	3)2580	6)3348	5)3745
2.	8)6483	9)8174	4)2509	2)1983	7)3490
3.	3)2922	6)5277	5)4350	8)6543	9)6255

Perfect score: 15   My score: _____

Lesson 3   Division

Think about each digit in the quotient.

thousands

$$4\overline{)9}$$

$$\begin{array}{r} 2 \\ 4\overline{)9527} \\ 8000 \\ \hline 1527 \end{array}$$

hundreds

$$4\overline{)15}$$

$$\begin{array}{r} 23 \\ 4\overline{)9527} \\ 8000 \\ \hline 1527 \\ 1200 \\ \hline 327 \end{array}$$

tens

$$4\overline{)32}$$

$$\begin{array}{r} 238 \\ 4\overline{)9527} \\ 8000 \\ \hline 1527 \\ 1200 \\ \hline 327 \\ 320 \\ \hline 7 \end{array}$$

ones

$$4\overline{)7}$$

$$\begin{array}{r} 2381 \;\; r3 \\ 4\overline{)9527} \\ 8000 \\ \hline 1527 \\ 1200 \\ \hline 327 \\ 320 \\ \hline 7 \\ 4 \\ \hline 3 \end{array}$$

Divide.

	a	*b*	*c*	*d*	*e*
1.	$6\overline{)6684}$	$3\overline{)9642}$	$8\overline{)8032}$	$5\overline{)7845}$	$9\overline{)9819}$
2.	$5\overline{)5758}$	$4\overline{)4287}$	$7\overline{)9466}$	$6\overline{)8755}$	$4\overline{)9979}$

Perfect score: 10   My score: _____

Problem Solving

Solve each problem.

1. A trip is 956 kilometers long. The same distance will be driven on each of 2 days. How many kilometers will be driven on each day?

_____ kilometers will be driven each day.

2. A pony-express trip of 1,230 miles took 5 days. The rider went the same number of miles each day. How many miles did the rider go each day?

The rider went _____ miles each day.

3. The Thompsons' car cost $5,124. They paid the same amount in each of 3 years. How much did they pay each year?

_____ was paid each year.

4. A tank contains 555 liters of oil. Nine liters of oil are used each day. How many days will the supply last? How many liters will be left over?

The supply will last _____ days.

_____ liters will be left over.

5. At St. Thomas School there are 4 grades and 4,196 pupils. There is the same number of pupils in each grade. How many pupils are in each grade?

_____ pupils are in each grade.

6. There are 5,280 feet in a mile and 3 feet in a yard. How many yards are there in a mile?

_____ yards are in a mile.

7. Each box holds 8 cans. There are 9,539 cans to be packed. How many boxes will be filled? How many cans will be left over?

_____ boxes will be filled.

_____ cans will be left over.

1.	
2.	**3.**
4.	**5.**
6.	**7.**

Perfect score: 9   My score: _____

108

CHAPTER 10 TEST

Divide.

	a	*b*	*c*	*d*	*e*
1.	3)216	6)481	4)2284	2)1256	6)1848
2.	7)5478	8)3645	5)4378	8)6245	4)3675
3.	9)9819	7)8442	5)7605	5)5256	7)9184
4.	3)8252	6)8755	8)9147	9)9171	3)8678

10

Perfect score: 20   My score: _____

PRE-TEST—Multiplication and Division

Multiply or divide. Check each answer.

	a	*b*	*c*
1.	213 ×2	2202 ×4	3123 ×3
2.	787 ×8	1654 ×5	359 ×6
3.	2)126	3)679	8)964
4.	5)4507	9)8722	6)6726
5.	4)1345	7)1234	5)6752

Perfect score: 15   My score: _____

Lesson 1   Multiplication and Division

$$
\begin{array}{l}
19 \dashrightarrow 19 \\
\times 5 \rightarrow 5\overline{)95} \\
\overline{95} \dashrightarrow
\end{array}
\qquad
\begin{array}{l}
507 \\
\times 8 \\
\overline{4056} \quad \text{so} \quad 8\overline{)4056}
\end{array}
\qquad
\begin{array}{l}
13 \dashrightarrow 13 \\
7\overline{)91} \quad \times 7 \\
\underline{70} \rightarrow 91 \\
\overline{21} \\
\underline{21} \\
\overline{0}
\end{array}
\qquad
\begin{array}{l}
272 \qquad 272 \\
7\overline{)1904} \quad \text{so} \quad \times 7 \\
\underline{}
\end{array}
$$

Complete the following.

	a		*b*

1. $\begin{array}{l} 117 \\ \times 5 \\ \overline{585} \end{array}$ so $5\overline{)585}.$ $\qquad$ $3\overline{)285}$ so $\begin{array}{l}95\\ \times 3\end{array}.$

2. $\begin{array}{l} 219 \\ \times 7 \\ \overline{1533} \end{array}$ so $7\overline{)1533}.$ $\qquad$ $9\overline{)1827}$ so $\begin{array}{l}203\\ \times 9\end{array}.$

Multiply or divide.

	a	*b*	*c*	*d*

3. $\begin{array}{r} 2\,7 \\ \times\,5 \\ \hline \end{array}$ $\qquad 5\overline{)1\ 3\ 5}$ $\qquad \begin{array}{r} 1\,2\,3 \\ \times\,8 \\ \hline \end{array}$ $\qquad 8\overline{)9\ 8\ 4}$

4. $\begin{array}{r} 7\,2\,8 \\ \times\,6 \\ \hline \end{array}$ $\qquad 6\overline{)4\ 3\ 6\ 8}$ $\qquad 4\overline{)8\ 3\ 2\ 4}$ $\qquad \begin{array}{r} 2\,0\,8\,1 \\ \times\,4 \\ \hline \end{array}$

5. $9\overline{)2\ 2\ 1\ 4}$ $\qquad \begin{array}{r} 2\,4\,6 \\ \times\,9 \\ \hline \end{array}$ $\qquad 3\overline{)5\ 6\ 1}$ $\qquad \begin{array}{r} 1\,8\,7 \\ \times\,3 \\ \hline \end{array}$

Perfect score: 16   My score: _____

Problem Solving

Solve each problem.

1. Charles worked 80 problems in 5 days. He worked the same number of problems each day. How many problems did he work each day?

He worked _____ problems each day.

2. There are 9 rows of tiles on a floor. There are 18 tiles in each row. How many tiles are there on the floor?

_____ tiles are on the floor.

3. An 8-story apartment building is 112 feet high. Each story is the same height. What is the height of each story?

Each story is _____ feet high.

4. A sheet of plywood weighs 21 kilograms. What would be the weight of 8 sheets?

The weight would be _____ kilograms.

5. A company made 4,325 cars in 5 days. The same number of cars was made each day. How many cars were made each day?

_____ cars were made each day.

6. The seating capacity of a sports arena is 7,560. The seats are arranged in 6 sections of the same size. How many seats are there in each section?

_____ seats are in each section.

7. The school library has 8,096 books. The same number of books was stored along each of the 4 walls. How many books are along each wall?

_____ books are along each wall.

1.	
2.	**3.**
4.	**5.**
6.	**7.**

Perfect score: 7   My score: _____

112

Lesson 2   Multiplication

These should
be the same. – – – – – ► 321

321 ↙
×3
963

Check   3)963
900
63
60
3
3
0

To check 3 × 321 = 963,
divide 963 by 3.

These should
be the same. – – – – – ► 1243

1243 ↙
×4
4972

Check   4)4972
4000
972
800
172
160
12
12
0

To check 4 × 1243 = 4972,

divide _____ by ____.

Multiply. Check each answer.

	a	*b*	*c*
1.	2 3 1 ×3	6 7 8 ×7	9 7 5 ×8
2.	1 2 3 4 ×2	2 6 7 5 ×9	1 2 5 7 ×6
3.	2 3 0 2 ×4	7 5 8 2 ×5	3 5 4 3 ×7

Perfect score: 9   My score: _____

113

Problem Solving

Solve each problem. Check each answer.

1. Each of the 4 members of a relay team runs 440 yards. What is the total distance the team will run?

The team will run _____ yards.

2. Herman delivers 165 papers each day. How many papers does he deliver in a week?

He delivers _____ papers in a week.

3. There are 125 nails in a 1-pound pack. How many nails will be in a 5-pound pack?

_____ nails will be in a 5-pound pack.

4. It takes 1,200 trading stamps to fill a book. How many stamps will it take to fill 6 books?

It will take _____ stamps.

5. A contractor estimated that it would take 2,072 bricks to build each of the 4 walls of a new house. How many bricks would it take to build all 4 walls?

It would take _____ bricks.

6. There are 7 boxes on a truck. Each box weighs 650 kilograms. What is the weight of all the boxes?

The weight of all the boxes is _____ kilograms.

7. Each box in problem 6 has a value of $845. What is the value of all the boxes?

The value is $_____.

8. Miss Brooks travels 1,025 kilometers each week. How many kilometers will she travel in 6 weeks?

She will travel _____ kilometers.

1.	2.
3.	4.
5.	6.
7.	8.

Perfect score: 8   My score: _____

Lesson 3   Division

```
   442
3│1326                    Check    442
  1200                             × 3
  ─────  These should ─ ─ ─ ─→    1326
   126   be the same.
   120
  ─────
     6
     6
  ─────
     0
         To check 1326 ÷ 3 = 442,
         multiply 442 by 3.
```

```
   3453
5│17265                   Check    3453
  15000                            × 5
  ─────  These should ─ ─ ─→      17265
   2265  be the same.
   2000
  ─────
    265
    250
  ─────
     15  To check 17265 ÷ 5 = 3453,
     15
  ─────
      0  multiply 3453 by ____.
```

Divide. Check each answer.

	a	*b*	*c*
1.	3│2 4 6	5│6 7 5	9│9 8 1
2.	9│1 5 6 6	7│7 8 4 7	4│6 2 5 6
3.	3│1 7 1 0	7│1 4 3 5	2│9 8 5 8

Perfect score: 9   My score: _____

Problem Solving

Solve each problem. Check each answer.

1. A plane traveled 900 kilometers in 2 hours. The same distance was traveled each hour. How far did the plane travel each hour?

_____ kilometers were traveled each hour.

2. A lunchroom served 960 lunches in 3 hours. The same number of lunches was served each hour. How many lunches were served each hour?

_____ lunches were served each hour.

3. A company has 7,200 workers. There are 8 plants with the same number of workers at each plant. How many workers are at each plant?

_____ workers are at each plant.

4. A school ordered 2,688 books. The books will be delivered in 7 equal shipments. How many books will be in each shipment?

_____ books will be in each shipment.

5. 5,900 kilograms of freight was delivered in 5 loads. Each load had the same weight. Find the weight of each load.

The weight of each load was _____ kilograms.

6. There are 5,040 calories in 6 cups of roasted peanuts. How many calories are there in 1 cup of roasted peanuts?

_____ calories are in 1 cup of peanuts.

7. A refinery filled 8,652 oil cans in 3 hours. The same number of cans was filled each hour. How many cans were filled each hour?

_____ cans were filled each hour.

1.	
2.	3.
4.	5.
6.	7.

Perfect score: 7   My score: _____

116

Lesson 4   Division

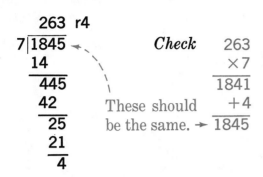

```
        263 r4
    7⟌1845 ⟵
       14
      ---
       445
        42
       ---
        25
        21
        --
         4
```

Check   263
             ×7

            1841
              +4

            1845

These should
be the same. ➝

To check the division,
multiply 263 by 7 and
then add the remainder 4.

Divide. Check each answer.

     a   *b*   *c*

1. 2⟌1 5 7   5⟌6 7 9   4⟌8 0 7

2. 3⟌1 3 4 9   6⟌7 8 5 5   7⟌9 4 6 3

3. 9⟌1 2 3 4   8⟌9 6 5 4   6⟌9 7 8 5

Perfect score: 9   My score: _____

Problem Solving

Solve each problem. Check each answer.

1. A marching band has 126 members. How many rows of 8 members each can be formed? How many members will be in the next row?

_____ rows can be formed.

_____ members will be in the next row.

2. Mr. Carpenter has 228 floor tiles. How many rows of 9 tiles each can he lay? How many tiles will be left over?

He can lay _____ rows of 9 tiles each.

_____ tiles will be left over.

3. 7 pirates want to share 1,006 coins so that each will get the same number of coins. How many coins will each pirate get? How many coins will be left over?

Each pirate will get _____ coins.

_____ coins will be left over.

4. There will be 1,012 people at a large banquet. Six people will sit at each table. How many tables will be filled? How many people will sit at a table that is not filled?

_____ tables will be filled.

_____ people will sit at a table that is not filled.

5. Last week 2,407 empty bottles were returned to a store. How many cartons of 8 bottles each could be filled? How many bottles would be left over?

_____ cartons could be filled.

_____ bottles would be left over.

| 1. |
| 2. |
| 3. |
| 4. |
| 5. |

Perfect score: 10   My score: _____

118

CHAPTER 11 TEST

Multiply or divide. Check each answer.

	a	*b*	*c*
1.	3 1 2 ×3	2 1 3 4 ×2	9 3 2 0 ×4
2.	6 3 6 ×7	1 2 5 4 ×6	2 3 4 1 ×8
3.	2⟌1 8 6	5⟌1 5 3 4	9⟌1 8 2 7
4.	4⟌4 7 7	6⟌6 7 1 4	8⟌9 6 4 2
5.	6⟌9 8 6	3⟌6 0 6 8	5⟌8 7 0 5

Perfect score: 15   My score: _____

PRE-TEST—Metric Measurement

Find the length of each line segment to the nearest centimeter (cm).
Then find the length of each line segment to the nearest millimeter (mm).

     a   *b*

1. _____ cm   _____ mm   ████████████████

2. _____ cm   _____ mm   ██████████████████████

3. _____ cm   _____ mm   ████████

Find the perimeter of each figure.

4.

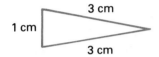

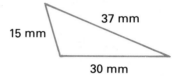

_____ centimeters   _____ millimeters

Complete the following.

        a   *b*

5. 9 centimeters = _____ millimeters   7 kiloliters = _____ liters

6. 4 meters = _____ centimeters   5 grams = _____ milligrams

7. 8 meters = _____ millimeters   13 meters = _____ millimeters

8. 6 liters = _____ milliliters   1 kilometer = _____ meters

9. 9 kilograms = _____ grams   29 liters = _____ milliliters

10. 12 meters = _____ centimeters   23 kiloliters = _____ liters

11. 1 gram = _____ milligrams   80 centimeters = _____ millimeters

12. 92 kilograms = _____ grams   16 kilometers = _____ meters

Perfect score: 24   My score: _____

Lesson 1 Centimeter

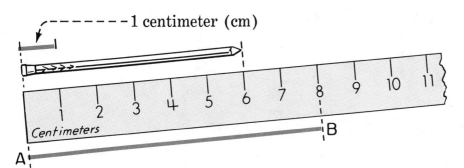

The nail is ___6___ centimeters long.   Line segment AB is _____ centimeters long.

Find the length of each picture to the nearest centimeter.

1. _____ cm

2. _____ cm

3. _____ cm

4. _____ cm

5. _____ cm

6. _____ cm

Use a ruler to draw a line segment for each measurement.

7. 5 cm

8. 8 cm

9. 3 cm

10. 6 cm

Perfect score: 10   My score: _____

Lesson 2   Millimeter

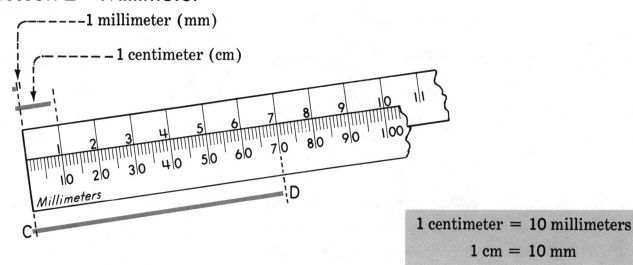

1 centimeter = 10 millimeters
1 cm = 10 mm

Line segment CD is _____7_____ centimeters or _____ millimeters long.

Find the length of each line segment to the nearest centimeter.
Then find the length of each line segment to the nearest millimeter.

      a   *b*

1. _____ cm   _____ mm

2. _____ cm   _____ mm

3. _____ cm   _____ mm

4. _____ cm   _____ mm

Find the length of each line segment to the nearest millimeter.

5. _____ mm

6. _____ mm

7. _____ mm

8. _____ mm

Use a ruler to draw a line segment for each measurement.

9. 50 mm

10. 80 mm

11. 25 mm

12. 55 mm

Perfect score: 16   My score: _____

Lesson 3   Perimeter

The distance around a figure is called its **perimeter**.

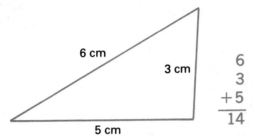

$$\begin{array}{r} 6 \\ 3 \\ +5 \\ \hline 14 \end{array}$$

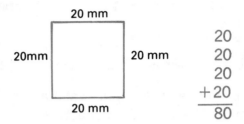

$$\begin{array}{r} 20 \\ 20 \\ 20 \\ +20 \\ \hline 80 \end{array}$$

The perimeter is ___14___ centimeters.   The perimeter is _____ millimeters.

Find the perimeter of each figure.

        a   *b*   *c*

1.

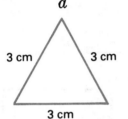

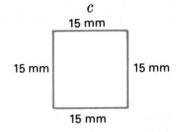

perimeter: _____ cm   *perimeter:* _____ mm   *perimeter:* _____ mm

Find the length of each side in centimeters. Then find the perimeter.

        a   *b*

2.

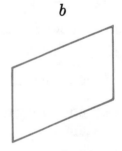

perimeter: _____ cm   *perimeter:* _____ cm

Find the length of each side in millimeters. Then find the perimeter.

3.

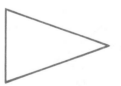

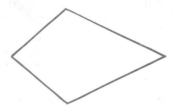

perimeter: _____ mm   *perimeter:* _____ mm

Perfect score: 7   My score: _____

Problem Solving

Solve each problem.

1. Find the perimeter of the rectangle to the nearest centimeter.

The perimeter is _____ centimeters.

2. Yvonne wants to glue yarn along the sides of the rectangle. She has 22 centimeters of yarn. How many centimeters of yarn will not be used?

_____ centimeters of yarn will not be used.

3. Find the perimeter of the blue square in centimeters. Do the same for the black square.

_____ centimeters is the perimeter of the blue square.

_____ centimeters is the perimeter of the black square.

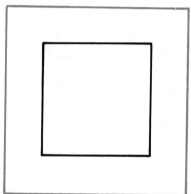

4. How much greater is the perimeter of the blue square than that of the black square?

The perimeter is _____ centimeters greater.

What is the combined distance around the two squares?

The combined distance is _____ centimeters.

Guess the perimeter of each of the following in centimeters. Then find each perimeter to the nearest centimeter.

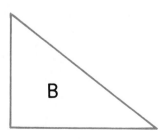

	Object	Guess	Perimeter
5.	rectangle A	_____ cm	_____ cm
6.	triangle B	_____ cm	_____ cm
7.	cover of this book	_____ cm	_____ cm
8.	cover of a dictionary	_____ cm	_____ cm
9.	top of a chalk box	_____ cm	_____ cm

Perfect score: 11   My score: _____

124

Lesson 4 Meter and Kilometer

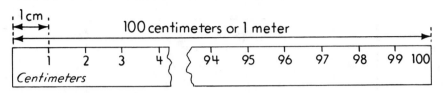

100 centimeters = 1 meter
100 cm = 1 m

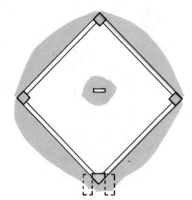

Suppose you run around a baseball diamond 9 times. You would run about 1 kilometer (km).

1000 meters = 1 kilometer
1000 m = 1 km

Find the length of each of the following to the nearest meter.

	Object	Length
1.	width of a door	_____ m
2.	height of a door	_____ m
3.	length of a chalkboard	_____ m
4.	height of a cabinet	_____ m

Solve each problem.

5. You and some classmates lay 5 of your math books like this. Find the length to the nearest meter.

The length is _____ meter.

6. Marcus lives 3 kilometers from school. How many meters is that?

The distance is _____ meters.

6.

7. Ms. Kahn can drive 87 kilometers in one hour. How many kilometers can she drive in 4 hours?

She can drive _____ kilometers in 4 hours.

7.

Perfect score: 7   My score: _____

Lesson 5   Units of Length

25 cm = ___?___ mm 1 cm = 10 mm ↓ ↓ 1 10 ×25 ×25 ‾‾‾‾ ‾‾‾‾ 25 250 ↓ ↓ 25 cm = __250__ mm	18 m = ___?___ cm 1 m = 100 cm ↓ ↓ 1 100 ×18 ×18 ‾‾‾‾ ‾‾‾‾ 18 1800 ↓ ↓ 18 m = __1800__ cm
9 m = ___?___ mm 1 m = 1000 mm ↓ ↓ 1 1000 ×9 ×9 ‾‾‾ ‾‾‾‾ 9 9000 ↓ 9 m = _____ mm	7 km = ___?___ m 1 km = 1000 m ↓ ↓ 1 1000 ×7 ×7 ‾‾‾ ‾‾‾‾ 7 7000 ↓ 7 km = _____ m

Complete the following.

	a	b
1.	9 cm = _____ mm	7 cm = _____ mm
2.	9 m = _____ cm	6 m = _____ cm
3.	9 m = _____ mm	4 m = _____ mm
4.	9 km = _____ m	5 km = _____ m
5.	16 m = _____ cm	8 m = _____ mm
6.	89 km = _____ m	46 m = _____ cm
7.	28 cm = _____ mm	18 km = _____ m
8.	13 m = _____ mm	42 cm = _____ mm
9.	16 m = _____ mm	10 m = _____ cm
10.	10 km = _____ m	25 m = _____ mm

Perfect score: 20   My score: _____

Lesson 6   Liter and Milliliter

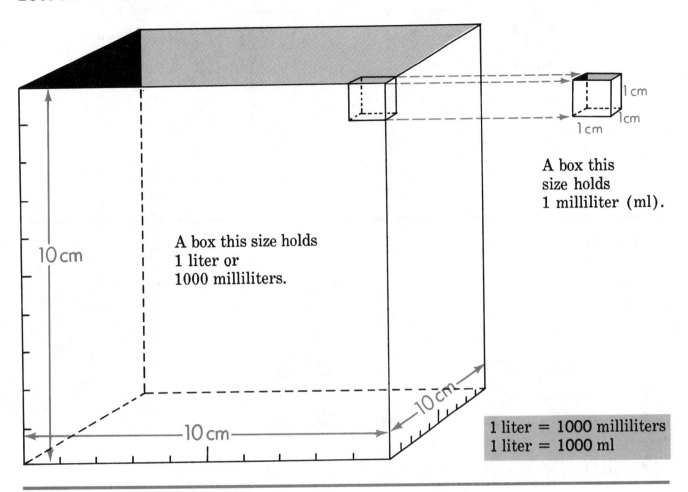

A box this size holds
1 liter or
1000 milliliters.

10 cm

10 cm

10 cm

A box this
size holds
1 milliliter (ml).

1 cm
1 cm
1 cm

1 liter = 1000 milliliters
1 liter = 1000 ml

Solve each problem.

1. How many milliliters does the carton hold?

The carton holds _____ milliliters.

2. Which container holds more liquid, the carton or the bottle?

The _____ holds more liquid.

1 liter   800 ml

3. A small milk carton holds 236 milliliters of milk. How many milliliters of milk are in 8 cartons?

There are _____ milliliters of milk.

4. Lyle can drive his car 11 kilometers per liter of gasoline. How far can he drive on 80 liters?

He can go _____ kilometers.

3.

4.

Perfect score: 4   My score: _____

Lesson 7   Liter and Milliliter

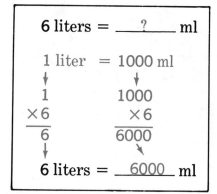

6 liters = ___?___ ml

1 liter = 1000 ml

1   1000
×6   ×6
6   6000

6 liters = __6000__ ml

13 liters = ___?___ ml

1 liter = 1000 ml

1   1000
×13   ×13

13 liters = _____ ml

Complete the following.

	a	b
1.	1 liter = _____ ml	2 liters = _____ ml
2.	3 liters = _____ ml	6 liters = _____ ml
3.	7 liters = _____ ml	9 liters = _____ ml
4.	10 liters = _____ ml	11 liters = _____ ml
5.	25 liters = _____ ml	30 liters = _____ ml

Solve each problem.

6. One tablespoon holds 15 milliliters. How many tablespoons of soup are in a 225-milliliter can?

There are _____ tablespoons of soup.

7. Barb used 8 liters of water when she washed her hands and face. How many milliliters of water did she use?

She used _____ milliliters of water.

There are 28 pupils in Barb's class. Suppose each pupil uses as much water as Barb. How many liters would be used?

_____ liters would be used.

6.

7.

Perfect score: 13   My score: _____

Lesson 8 Gram, Milligram, and Kilogram

A vitamin tablet
weighs about
100 milligrams (mg).

A dime weighs
about 2 grams.

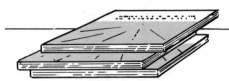

3 of your math
books weigh about
1 kilogram (kg).

1 gram = 1000 milligrams	1000 grams = 1 kilogram
1 g = 1000 mg	1000 g = 1 kg

Use the above diagrams to do problems **1–5.**

1. What is the weight of 2 vitamin tablets in milligrams?

The weight is _____ milligrams.

1.

2. Find the weight of 10 vitamin tablets in milligrams. Then find the weight in grams.

The weight is _____ milligrams or ____ gram.

2.

3. Find the weight in grams of a roll of 50 dimes.

The weight is _____ grams.

3.

4. What is the weight in grams of 10 rolls of dimes (500 dimes)? What is the weight in kilograms?

The weight is _____ grams or ____ kilogram.

4.

5. Find the weight in kilograms of a shipment of 30 math books. Then find the weight in grams.

It is ____ kilograms or _____ grams.

5.

Tell whether you would use *milligrams*, *grams*, or *kilograms* to weigh each object.

a	b	c
6. a nickel _____	a grain of sand _____	a bicycle _____
7. a pin _____	a new pencil _____	yourself _____

Perfect score: 14   My score: _____

Lesson 9   Weight

```
┌─────────────────────────────┐
│    7 g = ___?___ mg         │
│                             │
│    1 g = 1000 mg            │
│     ↓        ↓              │
│     1       1000           │
│    ×7        ×7            │
│    ───      ─────          │
│     7       7000           │
│     ↓        ↓              │
│    7 g = 7000 mg           │
└─────────────────────────────┘
```

```
┌─────────────────────────────┐
│   18 kg = ___?___ g         │
│                             │
│   1 kg = 1000 g            │
│    ↓        ↓              │
│    1       1000           │
│   ×18       ×18           │
│   ───      ─────          │
│    18      18000          │
│     ↓        ↓             │
│   18 kg = _____ g       │
└─────────────────────────────┘
```

Complete the following.

a   b

1. 5 kg = _____ g   9 g = _____ mg

2. 25 g = _____ mg   78 kg = _____ g

3. Tell the weight shown on each scale.

Al's Weight Last Year   Al's Weight This Year

4. How many kilograms did Al gain?

_____ kilograms   _____ kilograms   Al gained _____ kilograms.

5. Complete the table.

Saree's Breakfast

Food	Protein	Calcium
1 biscuit of shredded wheat	2 g	11 mg
1 cup of whole milk	8 g	291 mg
1 banana	1 g	10 mg
Total	_____ g	_____ mg

6. Give the amount of protein Saree had for breakfast in *milligrams*.

Saree had _____ milligrams of protein.

7. How many milligrams of calcium are in 4 cups of whole milk?

_____ milligrams of calcium are in 4 cups of whole milk.

Perfect score: 11   My score: _____

CHAPTER 12 TEST

Find the length of each line segment to the nearest centimeter.
Then find the length of each line segment to the nearest millimeter.

a   b

1. _____ cm   _____ mm   ━━━━━━━━━━

2. _____ cm   _____ mm   ━━━━━━━━━━━━━━━━

3. _____ cm   _____ mm   ━━━━━━━━━

Find the perimeter of each figure.

a   b   c

4.

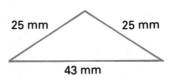

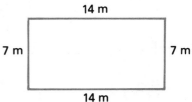

_____ centimeters   _____ millimeters   _____ meters

Complete the following.

a   b

5.   8 cm = _____ mm   2 kl = _____ liters

6.   2 m = _____ cm   6 g = _____ mg

7.   10 m = _____ mm   24 m = _____ mm

8.   3 liters = _____ ml   10 km = _____ m

9.   6 kg = _____ g   38 liters = _____ ml

10.   16 m = _____ cm   25 kl = _____ liters

11.   8 g = _____ mg   90 cm = _____ mm

12.   50 kg = _____ g   87 km = _____ m

Perfect score: 25   My score: _____

PRE-TEST—Measurement

1. Find the length of the line segment to the nearest $\frac{1}{2}$ inch.

_____ inches ▬▬▬▬▬▬▬▬▬▬▬▬

2. Find the length of the line segment to the nearest $\frac{1}{4}$ inch.

_____ inches ▬▬▬▬▬▬▬▬▬

Complete the following.

a   b

3. 3 pounds = _____ ounces   3 tons = _____ pounds

4. 3 minutes = _____ seconds   2 hours = _____ minutes

5. 4 days = _____ hours   6 yards = _____ feet

6. 3 yards = _____ feet   4 yards = _____ inches

7. 6 feet = _____ inches   6 feet = _____ yards

8. 8 quarts = _____ pints   4 gallons = _____ quarts

9. 6 pints = _____ cups   8 pints = _____ quarts

10. 16 quarts = _____ gallons   10 cups = _____ pints

Find the perimeter of each figure.

a   b

11.

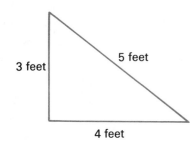
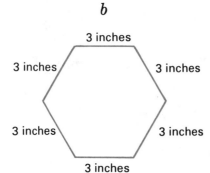

_____ feet   _____ inches

Perfect score: 20   My score: _____

Lesson 1 $\frac{1}{2}$ Inch

NAME _____

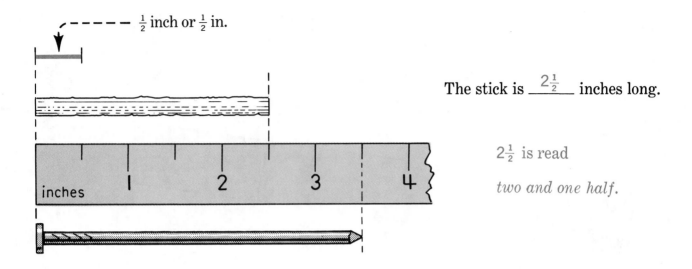

$\frac{1}{2}$ inch or $\frac{1}{2}$ in.

inches | 1 | 2 | 3 | 4

The stick is __$2\frac{1}{2}$__ inches long.

$2\frac{1}{2}$ is read

two and one half.

The nail is _____ inches long.

Find the length of each picture to the nearest $\frac{1}{2}$ inch.

1. _____ in.

2. _____ in.

3. _____ in.

4. _____ in.

5. _____ in.

6. _____ in.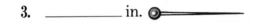

Use a ruler to draw a line segment for each measurement.

7. $1\frac{1}{2}$ in.

8. $3\frac{1}{2}$ in.

9. $4\frac{1}{2}$ in.

10. 5 in.

Perfect score: 10   My score: _____

133

Lesson 2 $\frac{1}{4}$ Inch

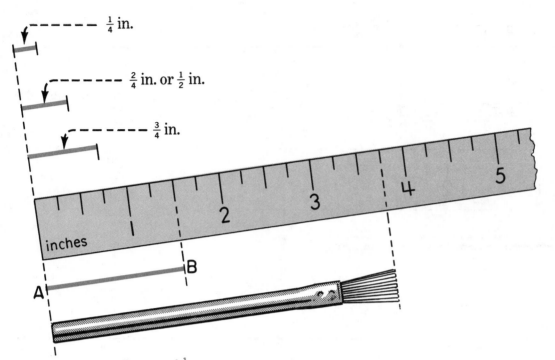

Line segment AB is __$1\frac{1}{2}$__ inches long.   The brush is _____ inches long.

Find the length of each picture to the nearest $\frac{1}{4}$ inch.

1. _____ in.

2. _____ in.

3. _____ in.

4. _____ in.

5. _____ in.

Use a ruler to draw a line segment for each measurement.

6. $2\frac{1}{4}$ in.

7. $1\frac{1}{2}$ in.

8. $4\frac{3}{4}$ in.

Perfect score: 8   My score: _____

Lesson 3 Units of Length

6 ft = _____?_____ yd

3 ft = 1 yd

$$\begin{array}{r} 2 \\ 3\overline{)6} \end{array}$$

6 ft = ___2___ yd

1 foot (ft) = 12 in.

1 yard (yd) = 3 ft or 36 in.

1 mile (mi) = 5280 ft

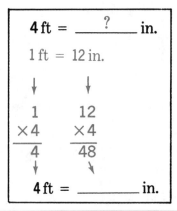

4 ft = _____?_____ in.

1 ft = 12 in.

$$\begin{array}{cc} 1 & 12 \\ \times 4 & \times 4 \\ \hline 4 & 48 \end{array}$$

4 ft = _____ in.

Complete the following.

	a	*b*
1.	3 ft = _____ in.	12 ft = _____ yd
2.	2 yd = _____ in.	5 yd = _____ in.
3.	5 ft = _____ in.	7 ft = _____ in.
4.	2 mi = _____ ft	6 yd = _____ in.
5.	15 ft = _____ yd	27 ft = _____ yd
6.	7 yd = _____ ft	9 yd = _____ ft
7.	9 ft = _____ in.	5 mi = _____ ft
8.	15 yd = _____ ft	75 ft = _____ yd
9.	60 ft = _____ yd	7 yd = _____ in.
10.	3 yd = _____ in.	300 ft = _____ yd
11.	8 ft = _____ in.	9 yd = _____ in.
12.	10 yd = _____ ft	3 mi = _____ ft

Perfect score: 24   My score: _____

Problem Solving

Solve each problem.

1. Mr. Jefferson is 6 feet tall. What is his height in inches?

His height is _____ inches.

2. In baseball the distance between home plate and first base is 90 feet. What is this distance in yards?

The distance is _____ yards.

3. Jeromy has 150 yards of kite string. How many feet of kite string does he have?

He has _____ feet of kite string.

4. A trench is 2 yards deep. What is the depth of the trench in inches?

The trench is _____ inches deep.

5. There are 5,280 feet in a mile. How many yards are there in a mile?

There are _____ yards in a mile.

6. One of the pro quarterbacks can throw a football 60 yards. How many feet can he throw the football?

He can throw the football _____ feet.

7. Marcena has 8 feet of ribbon. How many inches of ribbon does she have?

She has _____ inches of ribbon.

8. A rope is 3 yards long. What is the length of the rope in inches?

The rope is _____ inches long.

9. A certain car is 6 feet wide. What is the width of the car in inches?

The car is _____ inches wide.

1.	
2.	3.
4.	5.
6.	7.
8.	9.

Perfect score: 9   My score: _____

136

Lesson 4 Perimeter

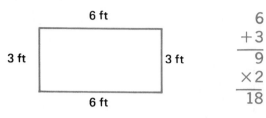

$$\begin{array}{r} 6 \\ +3 \\ \hline 9 \\ \times 2 \\ \hline 18 \end{array}$$

The perimeter is ____18____ feet.

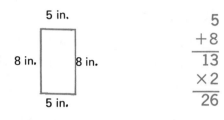

$$\begin{array}{r} 5 \\ +8 \\ \hline 13 \\ \times 2 \\ \hline 26 \end{array}$$

The perimeter is _____ inches.

To find the perimeter of a rectangle,
(1) add the measures of the length and width and
(2) multiply that sum by 2.

Find the perimeter of each rectangle below.

	a	*b*	*c*

1.

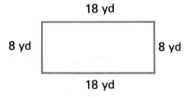

_____ feet

_____ inches

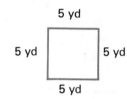

_____ yards

2.

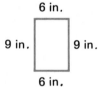

_____ inches

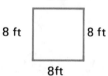

_____ feet

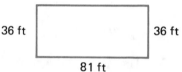

_____ feet

3.

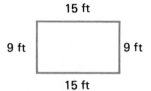

_____ yards

_____ inches

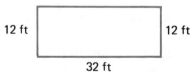

_____ feet

4.

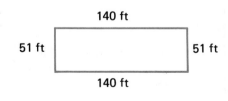

_____ feet

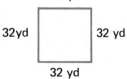

_____ yards

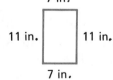

_____ inches

Perfect score: 12   My score: _____

137

Problem Solving

Solve each problem.

1. Mr. Champney's garden is shaped like a rectangle. The rectangle is 24 feet long and 16 feet wide. What is the perimeter of his garden?

The perimeter of his garden is _____ feet.

2. A rectangular desk top is 24 inches long and 16 inches wide. What is the perimeter of the desk top?

The perimeter of the desk top is _____ inches.

3. A flower garden is shaped like a rectangle. The length of the rectangle is 40 feet and the width is 30 feet. How many feet of edging will be needed to go around the garden?

_____ feet of edging will be needed.

4. A rectangular window pane is 28 inches long and 24 inches wide. What is the perimeter of the window pane?

The perimeter is _____ inches.

5. Mrs. Richardson has a rectangular-shaped mirror which is 4 feet long and 3 feet wide. How many feet of ribbon will she need to go around the edges of the mirror?

She will need _____ feet of ribbon.

6. A rectangular picture frame is 32 inches long and 24 inches wide. What is the perimeter of the picture frame?

The perimeter is _____ inches.

7. A football field is shaped like a rectangle. The length of the field is 360 feet and the width is 160 feet. What is the perimeter of a football field?

The perimeter is _____ feet.

1.	
2.	**3.**
4.	**5.**
6.	**7.**

Perfect score: 7   My score: _____

138

Lesson 5   Capacity

1 pint (pt) = 2 cups

1 quart (qt) = 2 pt

1 gallon (gal) = 4 qt

6 qt = ___?___ pt

1 qt = 2 pt

1   2
×6   ×6
‾‾6   ‾‾12

6 qt = _____ pt

12 qt = ___?___ gal

4 qt = 1 gal

$4\overline{)12}\;^{3}$

12 qt = _____ gal

Complete the following.

	a	b
1.	6 cups = _____ pt	12 qt = _____ pt
2.	4 pt = _____ qt	8 gal = _____ qt
3.	8 qt = _____ gal	6 pt = _____ cups
4.	8 pt = _____ cups	24 qt = _____ gal
5.	10 qt = _____ pt	18 pt = _____ qt
6.	5 gal = _____ qt	10 cups = _____ pt
7.	10 gal = _____ qt	32 qt = _____ gal
8.	12 pt = _____ qt	18 cups = _____ pt
9.	10 pt = _____ cups	32 pt = _____ qt
10.	28 qt = _____ gal	12 gal = _____ qt
11.	16 qt = _____ pt	28 qt = _____ pt
12.	8 cups = _____ pt	16 pt = _____ cups

Perfect score: 24   My score: _____

139

Problem Solving

Solve each problem.

1. There are 6 pints of lemonade in a picnic cooler. How many 1-cup containers can be filled by using the lemonade in the cooler?

_____ containers can be filled.

2. The cooling system on a car holds 16 quarts. How many gallons does it hold?

It holds _____ gallons.

3. In problem 2, how many pints does the cooling system hold?

It holds _____ pints.

4. There were 376 quarts of milk delivered to the store. How many gallons of milk was this?

It was _____ gallons of milk.

5. How many quarts of water would be needed to fill a 10-gallon aquarium?

_____ quarts would be needed.

6. The lunchroom served 168 pints of milk at lunch. How many quarts of milk was this?

It was _____ quarts of milk.

7. There are 12 cups of liquid in a container. How many 1-pint jars can be filled by using the liquid in the container?

_____ jars can be filled.

8. There are 6 pints of bleach in a container. How many quarts of bleach are in the container?

There are _____ quarts of
    bleach in the container.

1.	2.
3.	**4.**
5.	**6.**
7.	**8.**

Perfect score: 8   My score: _____

140

Lesson 6   Weight and Time

1 pound (lb) = 16 ounces (oz)
1 ton (T) = 2000 lb

1 minute (min) = 60 seconds (sec)
1 hour = 60 min
1 day = 24 hours

5 lb = _____?_____ oz

$$1 \text{ lb} = 16 \text{ oz}$$
$$\downarrow \qquad \downarrow$$
$$1 \qquad 16$$
$$\times 5 \qquad \times 5$$
$$\overline{5} \qquad \overline{80}$$
$$\downarrow \qquad \downarrow$$

5 lb = _____80_____ oz

3 min = _____?_____ sec

$$1 \text{ min} = 60 \text{ sec}$$
$$\downarrow \qquad \downarrow$$
$$1 \qquad 60$$
$$\times 3 \qquad \times 3$$
$$\overline{3} \qquad \overline{180}$$
$$\downarrow \qquad \downarrow$$

3 min = _____ sec

Complete the following.

	a	*b*
1.	2 lb = _____ oz	6 T = _____ lb
2.	2 T = _____ lb	4 lb = _____ oz
3.	7 lb = _____ oz	5 T = _____ lb
4.	2 hours = _____ min	8 min = _____ sec
5.	2 days = _____ hours	5 hours = _____ min
6.	5 min = _____ sec	3 days = _____ hours
7.	12 hours = _____ min	10 lb = _____ oz
8.	6 min = _____ sec	4 T = _____ lb
9.	5 days = _____ hours	10 min = _____ sec
10.	15 lb = _____ oz	24 hours = _____ min
11.	16 T = _____ lb	7 days = _____ hours
12.	4 min = _____ sec	9 lb = _____ oz

Perfect score: 24   My score: _____

Problem Solving

Solve each problem.

1. Mr. Werner bought a 5-pound roast. How many ounces did the roast weigh?

The roast weighed _____ ounces.

2. Susanne made a long-distance call that lasted 3 minutes. How many seconds did her call last?

Her call lasted _____ seconds.

3. A TV show started at 8 P.M. It lasted 60 minutes. What time was it over?

The show was over at _____ P.M.

4. A movie at the D.B.S. Theater lasted 3 hours. How many minutes did the movie last?

The movie lasted _____ minutes.

5. Elise weighs 80 pounds. How many ounces does she weigh?

Elise weighs _____ ounces.

6. The load limit on a small bridge is 8 tons. What is the load limit in pounds?

The load limit is _____ pounds.

7. A runner ran a mile in 4 minutes. What was the runner's time in seconds?

The runner's time was _____ seconds.

8. How many hours are there in a week?

_____ hours are in a week.

9. There are 30 tons of ore on a freight car. How many pounds of ore are on the freight car?

_____ pounds are on the freight car.

1.	
2.	**3.**
4.	**5.**
6.	**7.**
8.	**9.**

Perfect score: 9   My score: _____

Lesson 7 Problem Solving

NAME _____

Solve each problem.

1. To crochet the edging on a scarf you need 90 feet of yarn. How many yards is that?

You need _____ yards of yarn.

2. A board is 96 inches long. What is the length of the board in feet?

The board is _____ feet long.

3. Andre is going to put a fence around a garden shaped like a rectangle. The garden is 10 feet long and 15 feet wide. How many feet of fence will he need?

He will need _____ feet of fence.

4. A restaurant served 128 pints of milk in one day. How many quarts of milk was that?

The restaurant served _____ quarts of milk.

5. How many gallons of milk did the restaurant in 4 serve?

The restaurant served _____ gallons of milk.

6. A truck was loaded with 3 tons of cargo. How many pounds were on the truck?

There were _____ pounds of cargo on the truck.

7. Jackie's baby weighed 112 ounces when it was born. How many pounds did the baby weigh?

The baby weighed _____ pounds.

8. At halftime there were 180 seconds of commercials. How many minutes was that?

There were _____ minutes of commercials.

1.	2.
3.	4.
5.	6.
7.	8.

Perfect score: 8   My score: _____

CHAPTER 13 TEST

1. Find the length of the line segment to the nearest $\frac{1}{2}$ inch.

_____ inches ▬▬▬▬▬▬▬▬▬

2. Find the length of the line segment to the nearest $\frac{1}{4}$ inch.

_____ inches ▬▬▬▬▬▬▬▬▬▬

Complete the following.

	a		b

3.   5 lb = _____ oz   6 ft = _____ in.

4.   6 min = _____ sec   8 cups = _____ pt

5.   6 pt = _____ qt   12 ft = _____ yd

6.   3 yd = _____ in.   4 T = _____ lb

7.   9 qt = _____ pt   4 pt = _____ cups

8.   4 days = _____ hours   5 yd = _____ ft

9.   9 ft = _____ yd   5 gal = _____ qt

10.   12 qt = _____ gal   3 hours = _____ min

Find the perimeter of each figure.

        a   b

11.

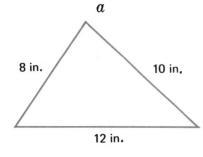

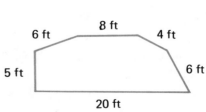

_____ inches   _____ feet

Perfect score: 20   My score: _____

144

NAME _____

Complete the following as indicated.

	a	b	c	d	e

1.

a:
```
  1 4
 +3
```
b:
```
  3 0
 +5 0
```
c:
```
  5 2
 +1 7
```
d:
```
  3 7
 +4
```
e:
```
  6 4
 +6 8
```

2.

a:
```
  1 3 2
 +8 8
```
b:
```
  2 3 5
 +1 6 9
```
c:
```
  2 7 6 3
 +1 4 2 8
```
d:
```
  3 2 4
  3 6 0
 +4 7
```
e:
```
  1 6 9 4
  3 4 7 0
  8 2 3 3
 +3 8 7 5
```

3.

a:
```
  2 8
 -5
```
b:
```
  8 0
 -2 0
```
c:
```
  3 7
 -2 1
```
d:
```
  4 2
 -1 6
```
e:
```
  1 2 7
 -6 5
```

4.

a:
```
  3 1 4
 -1 0 3
```
b:
```
  6 3 4
 -5 9 5
```
c:
```
  6 1 3 0
 -1 8 3
```
d:
```
  9 3 4 0
 -7 5 8 3
```
e:
```
  1 4 0 0 0
 -5 3 1 5
```

5.

a:
```
  2 0
 ×8
```
b:
```
  2 3
 ×3
```
c:
```
  1 5
 ×9
```
d:
```
  5 2 8
 ×4
```
e:
```
  2 0 4 6
 ×4
```

6.

a:
```
  4 3
 ×2 7
```
b:
```
  5 3 2
 ×8 5
```
c:
```
  2 4 6
 ×3 7 9
```
d:
```
  9 3 2
 ×8 1 6
```
e:
```
  5 3 9 1
 ×6 8
```

7.

a:
```
  5 2¢
 +3 4¢
```
b:
```
  $.6 2
 +.3 7
```
c:
```
  $3.2 1
 +1.6 9
```
d:
```
  $1 3.2 4
  1 7.5 3
     2.4 1
 +  8.4 9
```
e:
```
  $1 6.3 7
 -1 2.9 8
```

8.

a:
```
  9 3¢
 -6 4¢
```
b:
```
  $2 0.0 0
 -8.3 2
```
c:
```
  $.2 7
 ×7
```
d:
```
  $1.2 5
 ×1 3
```
e:
```
  $3.7 5
 ×2 5
```

Continued on the next page.

Test Ch. 1–7

TEST—Chapters 1–7 (Continued)

Record the temperature reading shown on each thermometer.

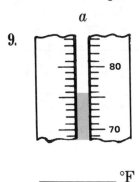   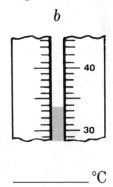      

9.

_____ °F   _____ °C   _____ °C   _____ °F

Solve each problem.

10. Last year Mona's bowling average was 117. This year her average is 132. How much has her bowling average improved over last year?

Her average has improved _____ points.

11. Ms. Wong got 3,243 votes. Mr. Williams got 2,952 votes. How many votes did the two people get in all?

The two people got _____ votes in all.

12. A machine makes 188 parts per hour. At that rate, how many parts can be made in 8 hours?

_____ items can be made in 8 hours.

13. There are 16 slices of cheese in each package. How many slices are in 18 packages?

There are _____ slices in 18 packages.

14. Anna earned $43.25 last week. She spent $17.18. How much does she have left?

Anna has $_____ left.

15. A factory made 650 machines per day. If the factory operated 266 days last year, how many machines were made?

_____ machines were made last year.

10.	11.
12.	**13.**
14.	**15.**

Perfect score: 50   My score: _____

FINAL TEST—Chapters 1–13

Complete the following as indicated.

	a	*b*	*c*	*d*	*e*

1.
$$\begin{array}{r} 21 \\ +5 \\ \hline \end{array} \qquad \begin{array}{r} 50 \\ +40 \\ \hline \end{array} \qquad \begin{array}{r} 31 \\ +67 \\ \hline \end{array} \qquad \begin{array}{r} 79 \\ +8 \\ \hline \end{array} \qquad \begin{array}{r} 58 \\ +95 \\ \hline \end{array}$$

2.
$$\begin{array}{r} 327 \\ +93 \\ \hline \end{array} \qquad \begin{array}{r} 527 \\ +457 \\ \hline \end{array} \qquad \begin{array}{r} 4626 \\ +1436 \\ \hline \end{array} \qquad \begin{array}{r} 538 \\ 256 \\ +57 \\ \hline \end{array} \qquad \begin{array}{r} 5134 \\ 1478 \\ 3163 \\ +5639 \\ \hline \end{array}$$

3.
$$\begin{array}{r} 67 \\ -2 \\ \hline \end{array} \qquad \begin{array}{r} 90 \\ -40 \\ \hline \end{array} \qquad \begin{array}{r} 65 \\ -45 \\ \hline \end{array} \qquad \begin{array}{r} 50 \\ -36 \\ \hline \end{array} \qquad \begin{array}{r} 241 \\ -89 \\ \hline \end{array}$$

4.
$$\begin{array}{r} 436 \\ -205 \\ \hline \end{array} \qquad \begin{array}{r} 625 \\ -356 \\ \hline \end{array} \qquad \begin{array}{r} 3027 \\ -549 \\ \hline \end{array} \qquad \begin{array}{r} 8423 \\ -6167 \\ \hline \end{array} \qquad \begin{array}{r} 17500 \\ -8447 \\ \hline \end{array}$$

5.
$$\begin{array}{r} 30 \\ \times 7 \\ \hline \end{array} \qquad \begin{array}{r} 21 \\ \times 4 \\ \hline \end{array} \qquad \begin{array}{r} 18 \\ \times 7 \\ \hline \end{array} \qquad \begin{array}{r} 479 \\ \times 5 \\ \hline \end{array} \qquad \begin{array}{r} 1624 \\ \times 3 \\ \hline \end{array}$$

6.
$$\begin{array}{r} 35 \\ \times 18 \\ \hline \end{array} \qquad \begin{array}{r} 178 \\ \times 84 \\ \hline \end{array} \qquad \begin{array}{r} 492 \\ \times 269 \\ \hline \end{array} \qquad \begin{array}{r} 808 \\ \times 735 \\ \hline \end{array} \qquad \begin{array}{r} 7346 \\ \times 49 \\ \hline \end{array}$$

7.
$$\begin{array}{r} \$.39 \\ +.45 \\ \hline \end{array} \qquad \begin{array}{r} \$2.35 \\ +3.59 \\ \hline \end{array} \qquad \begin{array}{r} \$25.19 \\ -13.59 \\ \hline \end{array} \qquad \begin{array}{r} \$.14 \\ \times 6 \\ \hline \end{array} \qquad \begin{array}{r} \$8.70 \\ \times 24 \\ \hline \end{array}$$

Final Test

Continued on the next page.

Final Test (Continued)

Divide.

	a	*b*	*c*	*d*

8. 8) 4 8   6) 3 0   9) 5 4   7) 4 2

9. 4) 3 0   8) 9 7   6) 3 0 8   5) 7 3 5

10. 3) 1 2 4 2   7) 6 3 4 9   6) 7 5 2 4   8) 9 9 4 7

Multiply or divide. Check each answer.

	a	*b*	*c*

11.
```
  3 2 6          2 9 8
   ×7     8)1 4 7 9    ×9
```

Complete each of the following.

	a	*b*

12. 5 cm = _____ mm   6 km = _____ m

13. 9 kg = _____ g   50 liters = _____ ml

14. 17 g = _____ mg   8 kl = _____ liters

15. 6 m = _____ mm   12 m = _____ cm

Continued on the next page.

Final Test (Continued)

Complete the following.

	a		b

16. 2 ft = _____ in.   3 lb = _____ oz

17. 6 qt = _____ pt   12 gal = _____ qt

18. 5 T = _____ lb   3 yd = _____ ft

19. 10 cups = _____ pt   2 days = _____ hours

20. 4 min = _____ sec   6 hours = _____ min

Find the perimeter of each figure.

                                 a   b

21.

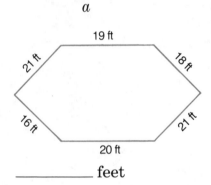

_____ feet

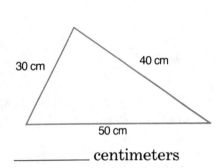

_____ centimeters

Solve each problem.

22. Jessica has saved $6,520 for a new car. Her goal is $9,780. How much more must she save?

She needs to save $_____ more.

22.

23.

23. There are 125 sheets of paper in a pack of paper. How many sheets are in 6 packs?

There are _____ sheets in 6 packs.

24. Andy bought 9 pounds of meat. The meat cost $27. What was the price for each pound?

The price for each pound was $_____.

24.

Continued on the next page.

Final Test (Continued)

Solve each problem.

25. A factory can make 128 whizzers in 8 hours. The same number of whizzers are made each hour. How many whizzers do they make each hour?

They make _____ whizzers each hour.

26. Adam earned $490 each week at his job. He also earned $48 each week from a part-time job. How much did Adam earn in all each week?

Adam earned $_____ in all each week.

27. A gasoline station sells an average of 847 gallons of gasoline per day. How many gallons will be sold in 365 days?

_____ gallons will be sold.

28. It takes 6 minutes to make each gizmo. How many gizmos can be made in 8 hours (480 minutes)?

_____ gizmos can be made.

29. Beverly drove 18,322 miles last year. Murray drove 9,785 miles last year. How many more miles did Beverly drive than Murray last year?

Beverly drove _____ more miles than Murray.

30. There are 1,455 items to be packed. The items are to be packed 8 to a box. How many boxes can be filled? How many items will be left over?

_____ boxes will be filled.

_____ items will be left over.

25.	26.
27.	**28.**
29.	**30.**

Perfect score: 80   My score: _____

Answers
Math - Grade 4
(Answers for Pre-Tests and Tests are given on pages 157–160.)

Page 3

	a	b	c	d	e	f	g	h
1.	8	7	9	9	7	8	9	10
2.	8	9	5	6	7	8	9	8
3.	5	7	4	9	9	5	6	7
4.	16	13	13	11	13	10	13	17
5.	14	14	11	11	10	12	12	13
6.	15	13	14	12	17	11	12	11
7.	10	10	12	12	11	12	16	18

Page 4

	a	b	c	d	e	f	g	h
1.	1	2	3	1	8	4	1	6
2.	2	1	7	6	4	0	3	3
3.	7	8	9	7	5	1	9	2
4.	6	3	8	6	8	8	7	4
5.	7	9	8	6	3	8	4	4
6.	5	9	7	6	9	7	9	7
7.	4	6	2	9	5	5	8	6

Page 5

	a	b	c	d	e	f
1.	27	32	43	19	27	58
2.	23	44	66	79	69	99
3.	69	18	19	29	39	39
4.	48	69	37	47	48	76
5.	11	31	54	21	42	91
6.	92	43	13	62	32	50
7.	85	24	42	73	86	63
8.	78	66	37	83	54	75

Page 6

1. 27 ; 6 ; 21   3. 23 ; 5 ; 28   5. 12
2. 8 ; 11 ; 19   4. 12 ; 6 ; 18

Page 7

	a	b	c	d	e	f
1.	6	60	7	70	9	90
2.	9	90	7	70	9	90
3.	90	80	80	80	80	60
4.	58	89	67	87	68	86
5.	36	97	77	98	95	84
6.	94	86	97	98	85	95
7.	99	89	88	89	89	98

Page 8

1. 14 ; 15 ; 29   3. 31 ; 28 ; 59   5. 38
2. 42 ; 46 ; 88   4. 43 ; 25 ; 68

Page 9

	a	b	c	d	e	f
1.	2	20	3	30	3	30
2.	5	50	2	20	4	40
3.	10	20	40	50	20	40
4.	11	10	28	12	10	12
5.	29	33	32	23	4	35
6.	2	43	40	55	15	46
7.	65	64	64	71	27	28

Page 10

1. 39 ; 23 ; 16   3. 47 ; 25 ; 22   5. 13
2. 28 ; 21 ; 7   4. 26 ; 16 ; 10

Page 13

	a	b	c	d	e	f	g	h
1.	13	12	14	18	17	12	17	14
2.	10	13	13	11	15	15	13	13
3.	13	16	16	14	11	11	11	11
4.	13	16	16	13	17	18	16	18
5.	18	17	15	17	16	15	15	10
6.	17	20	19	19	16	18	18	26

Page 14

1. 2 ; 3 ; 6 ; 11   3. 16   5. 11
2. 5 ; 4 ; 6 ; 15   4. 15   6. 13

Page 15

	a	b	c	d	e	f
1.	26	22	41	31	40	61
2.	31	71	65	42	42	73
3.	60	93	81	92	84	92
4.	71	41	41	50	43	72
5.	83	86	91	91	73	91
6.	73	57	74	77	43	46
7.	83	91	81	65	87	80

Page 16

1. 16 ; 9 ; 25   3. 63
2. 23   4. 92

Page 17

	a	b	c	d	e	f
1.	125	157	129	129	116	117
2.	128	138	119	176	159	139
3.	101	131	160	150	132	131
4.	188	206	178	248	139	149
5.	180	222	225	218	144	164
6.	100	203	220	201	265	135

Page 18

1. 61 ; 54 ; 115   3. 171   5. 158
2. 76 ; 83 ; 159   4. 174   6. 168

Page 19

	a	b	c	d	e	f
1.	18	56	63	37	77	68
2.	6	49	56	15	27	44
3.	27	59	18	39	9	16
4.	126	609	146	816	119	609
5.	517	147	408	106	736	108
6.	509	659	869	739	709	309
7.	309	718	618	309	529	845

Page 20

1. 32 ; 19 ; 13   3. 42 ; 27 ; 15   5. 108
2. 92 ; 68 ; 24   4. 216

Page 21

	a	b	c	d	e	f
1.	93	51	51	87	71	55
2.	331	390	771	391	691	534
3.	887	287	529	578	279	785
4.	99	809	288	669	188	788
5.	578	78	601	239	663	773
6.	387	268	89	248	759	268
7.	258	377	158	89	175	182

Answers

Page 22

1. 125 ; 64 ; 61   3. 865 ; 95 ; 770   5. 177
2. 117 ; 86 ; 31   4. 32

Page 23

	a	b	c	d	e	f
1.	67	98	88	73	80	94
2.	108	129	118	103	151	193
3.	61	34	29	67	39	25
4.	62	63	91	18	88	89

Page 24

	a	b	c	d	e	f
1.	21	15	23	16	18	14
2.	44	56	71	56	75	56
3.	71	91	117	120	384	528
4.	82	78	128	159	215	872
5.	29	39	55	38	46	67
6.	18	57	49	33	43	18
7.	118	239	745	619	217	628
8.	152	632	173	553	372	272
9.	269	449	778	277	659	397

Page 27

	a	b	c	d	e	f
1.	757	861	805	1139	521	816
2.	1582	1192	1219	1427	1413	1444
3.	885	640	1593	1209	1662	1650
4.	714	628	285	674	474	
5.	962	433	651	809	825	
6.	685	771	685	259	855	
7.	468	799	700	389	121	

Page 28

1. add ; 221   3. subtract ; 49
2. subtract ; 766   4. add ; 1313

Page 29

	a	b	c	d	e
1.	4677	7863	5749	7087	13648
2.	7803	9683	15764	7046	15868
3.	16055	7217	13542	8152	18141
4.	2239	3081	5821	4018	5588
5.	2008	5471	4922	3286	3084
6.	9459	9181	8811	485	8779
7.	10900	8709	6630	6499	3218

Page 30

1. 47682 ; 9613 ; 38069   3. 5016   5. 5347
2. 47682 ; 870 ; 48552   4. 1387

Page 31

	a	b	c	d	e
1.	95	84	178	79	146
2.	167	762	578	1875	858
3.	7833	7495	15586	7326	16866
4.	13196	8893	22821	13756	14974
5.	9792	9666	9377	17679	8298
6.	7860	9644	16782	18213	18312

Page 32

1. 707 ; 791 ; 729 ; 2227
2. 1687   3. 6568   4. 19835   5. 25579

Page 33

	a	b	c	d	e
1.	853	562	615	1146	685
2.	1674	2756	6136	4910	5221
3.	7463	8958	6073	12367	7511
4.	80	592	824	15958	12067
5.	227	1587	19289	20724	19069
6.	253	518	183	217	297
7.	1109	3192	3414	3818	4046

Page 33 (continued)

8.	3316	3271	1613	4811	3639
9.	40727	27116	36362	51855	62845

Page 34

1. 1750 ; 1402 ; 3152   4. 2012   7. 1788
2. 17421   5. 46
3. 2180   6. 6866

Page 37

	a	b	c	d	e	f	g	h
1.	6	56	72	0	49	42	7	24
2.	81	12	42	9	8	48	64	21
3.	0	35	0	18	56	54	0	28
4.	30	27	28	18	36	32	48	18
5.	63	35	40	40	24	21	45	12
6.	14	32	54	63	36	30	72	16
7.	24	45	24	16	27	14	36	18

Page 38

1. 5 ; 8 ; 40   3. 4 ; 5 ; 20   5. 48
2. 2 ; 8 ; 16   4. 25

Page 39

	a	b	c	d	e	f
1.	9	90	8	80	6	60
2.	6	90	96	2	80	82
3.	99	99	36	28	93	39
4.	64	69	84	84	26	66
5.	24	55	66	33	42	66
6.	44	88	44	88	22	26
7.	46	88	48	63	62	77

Page 40

1. 12 ; 2 ; 24   2. 36   3. 48   4. 40

Page 41

	a	b	c	d	e	f
1.	56	92	75	96	95	74
2.	84	91	87	96	92	84
3.	159	305	148	486	497	248
4.	368	146	455	427	216	488
5.	657	170	354	336	201	456
6.	432	304	747	490	378	378
7.	415	444	873	464	518	196

Page 42

1. 12 ; 3 ; 36   3. 24 ; 9 ; 216   5. 276
2. 14 ; 6 ; 84   4. 576

Page 43

	a	b	c	d	e	f
1.	6	600	35	3500	48	4800
2.	72	800	872	84	2800	2884
3.	963	846	848	698	981	
4.	936	952	4842	2891	3368	
5.	5922	4302	4746	5816	2970	

Page 44

1. 915   3. 1052   5. 1302   7. 768
2. 875   4. 1824   6. 1015

Page 45

	a	b	c	d	e	f
1.	77	64	84	96	48	69
2.	91	75	94	78	96	75
3.	216	368	426	248	405	279
4.	264	336	402	648	399	245

	a	b	c	d	e
5.	428	693	840	896	690
6.	585	917	788	968	2196
7.	3648	3050	942	940	959
8.	3248	4291	1581	6039	3440
9.	6587	3864	6075	5832	3052

Page 46

1. 360	3. 450	5. 1080	7. 2540
2. 245	4. 90	6. 625	8. 2120

Page 49

	a	b	c	d	e	f
1.	39	390	86	860	840	690
2.	68	340	408	96	640	736
3.	882	1056	869	961	297	384
4.	759	946	616	736	294	1034

Page 50

1. 10 ; 21 ; 210   3. 720
2. 15 ; 32 ; 480

Page 51

	a	b	c	d	e	f
1.	455	4550	296	2960	2560	3750
2.	1083	1147	1152	989	1218	663
3.	7553	2666	2288	2016	1088	2132
4.	3724	9118	5355	3055	6474	7912

Page 52

1. 792	3. 675	5. 648	7. 1150
2. 1261	4. 192	6. 1040	8. 1176

Page 53

	a	b	c	d	e	f
1.	480	960	840	1120	4500	5880
2.	882	736	924	299	759	528
3.	1075	1479	850	896	1008	1248
4.	3834	3096	2016	3444	4233	3128
5.	4964	2436	4214	3478	2976	1652

Page 54

1. 825	3. 616	5. 1584	7. 224
2. 2400	4. 720	6. 1050	

Page 55

	a	b	c	d	e	f
1.	1845	18450	2912	29120	33300	39150
2.	528	2640	3168	969	19380	20349

	a	b	c	d	e
3.	4876	5076	5929	11877	9888
4.	26568	21266	27027	71904	39508

Page 56

1. 7500	3. 3000	5. 4620	7. 1770
2. 2700	4. 5184	6. 6900	8. 6670

Page 57

	a	b	c	d	e	f
1.	1134	3150	2457	1728	4704	6231
2.	1173	5655	3822	5518	1833	2808

	a	b	c	d	e
3.	17328	34020	15696	32528	49868
4.	24948	4004	39196	33375	37492
5.	3537	77771	15170	23040	6897

Page 58

1. 600	3. 192	5. 1200	7. 869
2. 6000	4. 4320	6. 980	8. 5248

Page 61

	a	b	c	d	e	f
1.	6	6000	8	8000	9	9000
2.	21	21000	40	40000	72	72000

	a	b	c	d	e
3.	6824	8072	5355	8193	32084
4.	7343	40856	9684	8755	48609
5.	34026	48663	33306	35170	54016

Page 62

1. 45000	3. 15456	5. 8455	7. 39150
2. 17500	4. 4268	6. 12204	

Page 63

	a	b	c	d
1.	8000	80000	15000	150000
2.	80000	90000	350000	540000
3.	64992	68904	72080	57850
4.	120736	166288	311112	280116
5.	619974	490230	775200	360122

Page 64

1. 187500	3. 567360	5. 536854	7. 109172
2. 73692	4. 16788	6. 20340	8. 29440

Page 65

	a	b	c	d
1.	6824	9381	40505	9684
2.	7842	32688	87309	56488
3.	7835	28441	28632	52554
4.	52923	337463	354322	280116
5.	101156	85877	112896	55292
6.	323646	725790	583488	617826

Page 66

1. 30672	3. 2112	5. 6168	7. 17232
2. 132912	4. 14940	6. 74620	

Page 67

	a	b	c	d	e
1.	639	63900	2268	226800	223200
2.	68373	53297	88920	175299	108711
3.	368300	284256	419019	310392	560048

Page 68

1. 12500	3. 12875	5. 778800
2. 16875	4. 349305	

Page 69

	a	b	c	d
1.	43239	17303	55296	46008
2.	102024	68526	94284	459774
3.	166608	198008	83916	117594
4.	487153	466446	176175	332898
5.	455300	459608	374250	388994

Page 70

1. 235248	3. 249600	5. 82125	7. 352225
2. 62720	4. 96200	6. 16132	

Page 73

	a	b	c	d
1.	12	26	40	81
2.	75	63	104	9
3.	67	103	78	47

Page 74

1. 100	3. 91	5. 39	7. 35
2. 180	4. 20	6. 9	8. 5

Page 75

	a	b	c
1.	5	.25	83
2.	.10	.50	4
3.	25	.75	29
4.	.50	.10	6
5.	85	.95	60
6.	1.00	.05	99
7.	4.08	25	
8.	7.63	3	
9.	3.09	5	
10.	6.19	9	

	a	b
11.	5.79	4 ; 19
12.	18.75	8 ; 69

Page 76

	a	b	c	d	e
1.	67¢	72¢	$1.19	$13.68	$108.97
2.	56¢	29¢	$ .74	$ 1.78	$ 28.79
3.	85¢	93¢	$1.25	$14.71	$ 44.41
4.	19¢	3¢	$2.07	$ 3.38	$ 30.89
5.	75¢	96¢	$2.37	$10.68	$204.24
6.	$1.59	$.58	$1.22	$ 4.52	$ 32.88

Page 77

	a	b	c	d	e
1.	$ .84	$ .65	$ .78	$ 6.88	$ 48.69
2.	$ 1.86	$ 3.64	$ 3.48	$ 9.88	$ 77.37
3.	$ 3.84	$ 9.89	$ 17.68	$ 161.64	$1050.00
4.	$1013.88	$1016.61	$126.48	$1359.93	$3342.24

Page 78

1. 2.39   3. 17.52   5. 359.56   7. 118.75
2. 50   4. 2.80   6. 5.75   8. 3.78

Page 81

	a	b	c	d	e	f
1.	5	3	7	5	6	7
2.	7	7	6	5	9	3
3.	8	4	6	4	5	7
4.	8	1	4	8	0	2
5.	8	0	4	9	3	9
6.	6	9	8	0	5	9
7.	4	3	6	1	2	2

Page 82

1. 24 ; 3 ; 8   3. 8   5. 8
2. 5   4. 4

Page 83

	a	b	c	d	e			a	b	c	d	e
1.	6	7	8	9		5.		3	1	6	8	4
2.	6	7	8	9		6.		4	8	0	2	9
3.	9	2	5	5	1	7.		4	7	4	7	9
4.	7	7	6	0	3							

Page 84

1. 24 ; 6 ; 4   3. 9   5. 9   7. 8
2. 28 ; 7 ; 4   4. 8   6. 5

Page 85

	a	b	c	d	e
1.	6	7	8	9	
2.	6	7	8	9	
3.	1	2	7	3	4
4.	5	8	0	3	6
5.	9	4	2	1	5
6.	7	0	8	6	9
7.	6	8	8	9	9

Page 86

1. 8   3. 6   5. 2   7. 9
2. 4   4. 6   6. 3   8. 7

Page 87

	a	b	c	d	e
1.	4	7	8	7	
2.	6	8	6	0	
3.	9	8	9	2	7
4.	3	5	4	4	5
5.	8	3	0	6	5
6.	6	2	4	2	3
7.	1	9	3	1	2
8.	0	4	2	0	6
9.	8	1	5	0	1
10.	6	5	4	3	8

Page 88

1. 9   3. 7   5. 9   7. 6
2. 5   4. 8   6. 9   8. 8

Page 91

	a	b	c	d	e
1.	5 r2	5 r7	8 r1	5 r1	6 r2
2.	8 r5	7 r4	9 r4	7 r2	9 r6
3.	9 r2	7 r3	7 r7	9 r1	5 r3

Page 92

1. 3 ; 3   3. 7 ; 5   5. 7 ; 4
2. 6 ; 3   4. 5 ; 3

Page 93

	a	b	c	d	e
1.	13	12	11	32	22
2.	11 r3	11 r3	21 r2	13 r5	12 r3
3.	13	10 r6	29	12 r1	23

Page 94

1. 11   3. 20   5. 12
2. 15 ; 4   4. 13 ; 5   6. 12 ; 4

Page 95

	a	b	c	d	e
1.	7 r2	9 r1	9 r4	9 r4	8 r5
2.	14	12	11	31	21
3.	17	16	12	14	10
4.	10 r4	11 r2	12 r3	42 r1	21 r2
5.	26 r1	13 r2	16 r1	23 r2	12 r5

Page 96

1. 8 ; 3   3. 11   5. 6
2. 16   4. 17 ; 2   6. 7 ; 7

Page 97

	a	b	c	d	e
1.	62	61	63	38	59
2.	60 r7	60 r1	61 r3	62 r2	85 r1
3.	91 r7	84	53 r2	54	94 r3

Page 98

1. 78   3. 83   5. 85
2. 75   4. 91

Page 99

	a	b	c	d	e
1.	234	118	203	117	126
2.	109 r4	132 r2	108 r7	493 r1	118 r3
3.	140 r2	316 r1	125	219 r3	124

Page 100

1. 120   3. 106 ; 2   5. 135   7. 150
2. 144   4. 121   6. 124 ; 4

Page 101

	a	b	c	d	e
1.	63	81	83	79	78
2.	94 r3	89 r1	84 r2	87 r3	89 r2
3.	231	408	214	185	113
4.	107 r6	132 r5	119 r1	128 r3	173 r4

Page 102

1. 105   3. 48 ; 3   5. 30   7. 73
2. 220   4. 115   6. 104 ; 3

Page 105

	a	b	c	d	e
1.	134	153	141	160 r5	241 r1
2.	92 r3	46 r7	98	50 r6	84 r2
3.	68	218	196 r2	81 r5	125 r2

Page 106

	a	b	c	d	e
1.	624	613	860	558	749
2.	810 r3	908 r2	627 r1	991 r1	498 r4
3.	974	879 r3	870	817 r7	695

Page 107

	a	b	c	d	e
1.	1114	3214	1004	1569	1091
2.	1151r3	1071r3	1352r2	1459r1	2494r3

Page 108

1. 478   3. $1708   5. 1049   7. 1192 ; 3
2. 246   4. 61 ; 6   6. 1760

Page 111

	a	b	c	d
1.	117	285		
2.	219	1827		
3.	135	27	984	123
4.	4368	728	2081	8324
5.	246	2214	187	561

Page 112

1. 16   3. 14   5. 865   7. 2024
2. 162   4. 168   6. 1260

Page 113

	a	b	c
1.	693	4746	7800
2.	2468	24075	7542
3.	9208	37910	24801

Page 114

1. 1760   3. 625   5. 8288   7. 5915
2. 1155   4. 7200   6. 4550   8. 6150

Page 115

	a	b	c
1.	82	135	109
2.	174	1121	1564
3.	570	205	4929

Page 116

1. 450   3. 900   5. 1180   7. 2884
2. 320   4. 384   6. 840

Page 117

	a	b	c
1.	78 r1	135 r4	201 r3
2.	449 r2	1309 r1	1351 r6
3.	137 r1	1206 r6	1630 r5

Page 118

1. 15 ; 6   3. 143 ; 5   5. 300 ; 7
2. 25 ; 3   4. 168 ; 4

Page 121

1. 5   3. 3   5. 4
2. 10   4. 7   6. 8
7-10. Have your teacher check your work.

Page 122

	a	b
1.	7	70	5. 65
2.	4	40	6. 45
3.	5	50	7. 75
4.	3	30	8. 35

9-12. Have your teacher check your work.

Page 123

	a	b	c
1.	9	60	60
2.	5	10	
3.	80	100	

Page 124

1. 18   4. 8 ; 32   7. guess ; 98
2. 4   5. guess ; 14   8. guess ; Answers
3. 20 ; 12   6. guess ; 12   9. guess ; may vary.

Page 125

1-4.   Answers may vary.
5. 1   6. 3000   7. 348

Page 126

	a	b		a	b
1.	90	70	6.	89000	4600
2.	900	600	7.	280	18000
3.	9000	4000	8.	13000	420
4.	9000	5000	9.	16000	1000
5.	1600	8000	10.	10000	25000

Page 127

1. 1000   3. 1888
2. carton   4. 880

Page 128

	a	b		a	b
1.	1000	2000	5.	25000	30000
2.	3000	6000	6.	3375	
3.	7000	9000	7.	8000 ; 224	
4.	10000	11000			

Page 129

1. 200   4. 1000 ; 1
2. 1000 ; 1   5. 10 ; 10000
3. 100

	a	b	c
6.	grams	milligrams	kilograms
7.	milligrams	grams	kilograms

Page 130

	a	b		
1.	5000	9000	5.	11 ; 312
2.	25000	78000	6.	11000
3.	34	38	7.	1164
4.	4			

Page 133

1. $1\frac{1}{2}$ 3. 1   5. $4\frac{1}{2}$
2. $2\frac{1}{2}$ 4. $3\frac{1}{2}$ 6. 3

7-10.   Have your teacher check your work.

Page 134

1. $4\frac{1}{4}$ 3. $2\frac{1}{2}$ 5. 3
2. $3\frac{3}{4}$ 4. $1\frac{3}{4}$

6-9.   Have your teacher check your work.

Page 135

	a	b		a	b
1.	36	4	7.	108	26400
2.	72	180	8.	45	25
3.	60	84	9.	20	252
4.	10560	216	10.	108	100
5.	5	9	11.	96	324
6.	21	27	12.	30	15840

Page 136

1. 72   3. 450   5. 1760   7. 96   9. 72
2. 30   4. 72   6. 180   8. 108

Page 137

	a	b	c		a	b	c
1.	48	26	20	3.	52	48	88
2.	30	32	234	4.	382	128	36

Answers Grade 4

Page 138

1. 80	3. 140	5. 14	7. 1040
2. 80	4. 104	6. 112	

Page 139

	a	b		a	b
1.	3	24	7.	40	8
2.	2	32	8.	6	9
3.	2	12	9.	20	16
4.	16	6	10.	7	48
5.	20	9	11.	32	56
6.	20	5	12.	4	32

Page 140

1. 12	3. 32	5. 40	7. 6
2. 4	4. 94	6. 84	8. 3

Page 141

	a	b
1.	32	12000
2.	4000	64

Page 141 continued

3.	112	10000
4.	120	480
5.	48	300
6.	300	72
7.	720	160
8.	360	8000
9.	120	600
10.	240	1440
11.	32000	168
12.	240	144

Page 142

1. 80	4. 180	7. 240
2. 180	5. 1280	8. 168
3. 9	6. 16000	9. 60000

Page 143

1. 30	4. 64	7. 7
2. 8	5. 16	8. 3
3. 50	6. 6000	

Photo Credits

Camerique, 68; Glencoe staff photo, 2, 16, 28, 40; Santa Fe Railway, 50; Dan Witt, 82

Page vii

	a	b	c	d	e	f	g	h
1.	12	1	9	5	7	3	14	12
2.	9	6	13	11	10	2	15	9
3.	13	2	6	10	12	15	8	4
4.	4	11	6	0	8	11	8	8
5.	10	6	7	16	10	12	14	10
6.	13	14	8	7	5	7	5	4
7.	11	18	15	13	9	15	13	8
8.	14	7	11	9	9	11	13	11
9.	3	10	16	11	12	10	17	9
10.	10	14	8	17	12	9	16	12

Page viii

	a	b	c	d	e	f	g	h
1.	3	13	11	10	1	9	10	6
2.	12	0	10	8	13	6	11	8
3.	11	4	11	2	9	11	15	9
4.	4	10	7	5	7	2	12	18
5.	12	5	11	12	12	3	9	10
6.	8	6	13	8	17	16	10	6
7.	12	15	8	7	14	13	13	5
8.	8	14	9	11	17	15	9	9
9.	12	16	7	14	10	15	7	14
10.	9	4	16	13	11	8	14	10

Page ix

	a	b	c	d	e	f	g	h
1.	4	4	0	8	5	5	0	5
2.	3	6	1	4	5	4	3	3
3.	3	1	6	3	4	6	3	9
4.	4	7	8	7	2	9	9	6
5.	2	2	2	6	5	5	9	2
6.	1	9	1	4	1	8	6	7
7.	0	5	4	6	0	7	0	9
8.	7	7	5	7	7	9	6	9
9.	7	8	6	8	0	8	8	3
10.	9	9	2	8	2	8	7	5

Page x

	a	b	c	d	e	f	g	h
1.	9	4	9	4	6	7	0	6
2.	9	4	5	0	8	8	7	1
3.	9	3	7	2	3	6	5	9
4.	8	1	4	0	4	1	5	7
5.	6	0	6	0	2	1	9	5
6.	8	3	3	0	9	1	3	7
7.	7	1	6	1	8	6	6	3
8.	7	2	5	5	7	2	4	9
9.	2	6	5	1	4	3	9	8
10.	8	0	8	3	8	2	4	5

Page xi

	a	b	c	d	e	f	g	h
1.	16	6	21	20	3	10	18	0
2.	5	0	1	24	0	9	18	0
3.	28	0	7	9	24	15	0	18
4.	8	27	12	63	0	32	7	4
5.	72	4	40	20	63	6	28	16
6.	25	48	32	35	24	15	54	54
7.	9	56	0	56	2	40	24	0
8.	40	14	64	81	45	30	72	35
9.	36	42	18	48	8	0	12	27
10.	21	0	16	30	36	12	49	36

Page xii

	a	b	c	d	e	f	g	h
1.	24	4	21	6	0	18	14	4
2.	9	0	12	2	10	0	28	0
3.	20	14	0	8	21	8	0	25
4.	27	16	1	15	35	36	72	9
5.	36	30	40	32	0	48	24	81
6.	3	18	18	42	72	35	12	0
7.	63	28	32	40	16	7	0	18
8.	48	12	0	27	5	49	30	16
9.	64	6	20	36	45	45	54	0
10.	24	42	63	56	24	15	8	54

Page xiii

	a	b	c	d	e	f	g
1.	2	1	5	3	3	1	8
2.	5	2	3	3	2	9	4
3.	2	2	7	6	6	8	0
4.	1	3	0	9	4	6	8
5.	4	6	2	5	3	4	7
6.	1	5	7	6	5	9	7
7.	2	3	8	9	4	9	0
8.	2	1	7	6	5	3	5
9.	5	8	3	8	0	4	1
10.	7	6	7	0	9	2	8
11.	8	0	9	9	7	1	7
12.	6	9	8	6	5	1	4

Page xiv

	a	b	c	d	e	f	g
1.	8	1	5	9	9	4	2
2.	7	5	1	0	2	6	8
3.	8	9	8	3	9	9	1
4.	0	1	3	0	3	0	9
5.	7	7	3	2	2	6	2
6.	7	7	8	4	4	3	4
7.	1	9	3	0	2	0	3
8.	0	6	6	4	8	5	4
9.	5	6	8	5	4	7	9
10.	5	5	9	3	7	6	5
11.	7	1	4	7	2	0	3
12.	4	1	8	5	6	6	8

Page 1

	a	b	c	d	e	f
1.	49	78	69	29	47	78
2.	47	57	49	37	29	48
3.	80	80	90	70	90	70
4.	75	64	27	45	66	85
5.	98	36	64	92	83	85
6.	21	51	42	21	42	72
7.	40	20	20	40	30	50
8.	43	3	34	55	50	44
9.	8	66	78	6	79	15

Page 2

1. 3 ; 2 ; 5   2. 9 ; 5 ; 4   3. 4 ; 3 ; 7

Page 11

	a	b	c	d	e	f
1.	29	29	58	59	49	29
2.	36	83	34	97	35	58
3.	60	70	80	60	50	80
4.	95	67	83	78	96	49
5.	96	88	98	87	68	99
6.	62	28	41	57	73	
7.	40	10	50	50	20	
8.	10	43	48	26	7	
9.	34	65	53	21	47	

Page 12

	a	b	c	d	e	f
1.	8	12	20	15	24	17
2.	35	51	74	70	55	56
3.	92	106	116	251	337	413
4.	113	212	186	196	181	186
5.	19	74	59	36	88	67
6.	15	39	26	59	47	58
7.	62	742	341	892	597	781
8.	288	455	549	785	699	379
9.	509	819	590	892	105	387

Page 25

	a	b	c	d	e	f
1.	43	55	67	53	72	90
2.	92	126	125	241	343	443
3.	57	29	19	55	14	19
4.	71	42	391	57	288	217
5.	118					

Page 26

	a	b	c	d	e	f
1.	835	677	579	741	372	787
2.	618	655	818	1356	1298	1396
3.	510	412	8213	4591	5929	9186
4.	65	64	857	1886	6733	8508
5.	76	878	7868	8887	7838	9928
6.	423	329	794	149	5115	4423
7.	4219	5452	5851	7139	7715	5271
8.	3225	4249	2381	2755	4548	7899
9.	66203	32415	74091	78232	67421	75708

Page 35

	a	b	c	d	e
1.	667	871	837	2148	1807
2.	4483	6558	6366	11848	5613
3.	73	578	8374	1331	13372
4.	218		472	831	1694
5.	4181		4845	7371	57555
6.	1800				
7.	13865				

Page 36

	a	b	c	d	e	f
1.	28	54	40	80	60	80
2.	46	84	96	55	63	44
3.	66	42	86	69	88	68
4.	85	81	90	92	84	96
5.	459	248	568	469	288	415
6.	800	600	800	484	248	936
7.	945	496	975	968	688	849
8.	2800	1236	3696	3591	2472	880
9.	5810	4386	966	5384	2345	3476

Page 47

	a	b	c	d	e	f
1.	60	60	50	80	90	90
2.	96	86	84	69	66	48
3.	84	76	84	80	92	96
4.	420	720	280	426	368	486
5.	576	567	342	273	245	260

	a	b	c	d	e
6.	864	448	864	565	721
7.	917	846	3555	2448	2796
8.	959	992	3591	2754	1668
9.	4689	7528	2268	2195	5142

Page 48

	a	b	c	d	e	f
1.	640	690	770	2560	3920	2430
2.	903	736	748	897	1008	1184
3.	2257	2173	3116	4144	3196	4104

	a	b	c	d	e
4.	8442	10240	11856	8908	12267
5.	38064	11289	35342	24282	56712

Page 59

	a	b	c	d	e
1.	903	299	714	850	828
2.	3213	3024	1449	6408	2184
3.	6594	3936	6946	7905	24887
4.	8424	28044	11544	23433	12267
5.	38874	41088	30345	26313	61236

Page 60

	a	b	c	d
1.	35000	8428	9753	9462
2.	35655	61216	50904	114614
3.	256184	622457	363216	364650
4.	96300	226800	99840	89424
5.	385776	327096	594864	383250

Page 71

	a	b	c	d
1.	8084	6642	38889	54064
2.	23506	29694	87507	118598
3.	373744	174988	386456	439482
4.	68688	30861	116202	111088
5.	97528	403004	298944	547566

Page 72

	a	b	c	d	e
1.	32	69	4	58	
2.	79¢	$1.07	85¢	$12.58	$ 88.33
3.	52¢	$ .59	$3.62	$ 3.78	$ 8.95
4.	$.96	$3.22	$4.86	$22.23	$436.25
5.	49.93				
6.	2.77				

Page 79

	a	b	c	d	e
1.	17	34	73	49	
2.	89¢	81¢	$ .64	$14.64	$72.52
3.	32¢	9¢	$ .17	$ 3.03	$ 6.63
4.	$.87	$2.76	$11.12	$56.28	$77.76
5.	70.03				

Page 80

	a	b	c	d	e		a	b	c	d	e
1.	1	4	4	2	7	6.	9	3	7	1	5
2.	0	9	2	8	7	7.	7	5	3	9	1
3.	4	2	8	1	5	8.	2	8	0	4	6
4.	7	3	6	9	0	9.	6	1	8	4	2
5.	8	6	0	4	2	10.	9	7	3	0	5

Page 89

	a	b	c	d	e
1.	2	3	0	8	9
2.	2	1	1	9	9
3.	6	8	6	2	1
4.	7	5	6	1	4
5.	7	6	0	3	7
6.	5	4	5	4	6
7.	7	7	2	1	3
8.	3	4	8	7	9
9.	6	3	9	5	0
10.	7	0	9	5	8

Page 90

	a	b	c	d	e
1.	8 r2	9 r2	8 r1	7 r6	8 r5
2.	14	12	31	12	16
3.	13 r2	11 r5	10 r7	14 r3	13 r6
4.	62	59	23	47 r2	76 r4
5.	123	210	482	139 r1	128 r2

Page 103

	a	b	c	d	e
1.	6 r2	6 r3	9 r5	11	43
2.	18	13	22 r2	21 r3	12 r2
3.	52	41	73	57	84
4.	90 r3	92 r2	85 r2	69 r6	58 r1
5.	109	223	136	210 r3	123 r2

Page 104

	a	b	c	d
1.	83	90 r2	611 r5	709 r3
2.	2102	1009	1306 r5	1344 r4
3.	634	820 r2	402 r4	196
4.	1130	1230 r2	1205	1184

Page 109

	a	b	c	d	e
1.	72	80r1	571	628	308
2.	782r4	455r5	875r3	780r5	918r3
3.	1091	1206	1521	1051r1	1312
4.	2750r2	1459r1	1143r3	1019	2892r2

Page 110

	a	b	c
1.	426	8808	9369
2.	6296	8270	2154
3.	63	226 r1	120 r4
4.	901 r2	969 r1	1121
5.	336 r1	176 r2	1350 r2

Page 119

	a	b	c
1.	936	4268	37280
2.	4452	7524	18728
3.	93	306 r4	203
4.	119 r1	1119	1205 r2
5.	164 r2	2022 r2	1741

Page 120

	a	b
1.	5	50
2.	7	70
3.	3	30
4.	7	82
5.	90	7000
6.	400	5000
7.	8000	13000
8.	6000	1000
9.	9000	29000
10.	1200	23000
11.	1000	800
12.	92000	16000

Page 131

	a	b	c
1.	4	40	
2.	9	90	
3.	6	60	
4.	8	93	42

Page 131 continued

	a	b		a	b
5.	80	2000	9.	6000	38000
6.	200	6000	10.	1600	25000
7.	10000	24000	11.	8000	900
8.	3000	10000	12.	50000	87000

Page 132

1. $3\frac{1}{2}$
2. $2\frac{3}{4}$

	a	b		a	b
3.	48	6000	8.	16	16
4.	180	120	9.	12	4
5.	96	18	10.	4	5
6.	9	144	11.	12	18
7.	72	2			

Page 144

1. $2\frac{1}{2}$
2. $3\frac{1}{4}$

	a	b		a	b
3.	80	72	8.	96	15
4.	360	4	9.	3	20
5.	3	4	10.	3	180
6.	108	8000	11.	30	49
7.	18	8			

Page 145

	a	b	c	d	e
1.	17	80	69	41	132
2.	220	404	4191	731	17272
3.	23	60	16	26	62
4.	211	39	5947	1757	8685
5.	160	69	135	2112	8184
6.	1161	45220	93234	760512	366588
7.	86¢	$.99	$4.90	$41.67	$3.39
8.	29¢	$11.68	$1.89	$16.25	$93.75

Page 146

	a	b	c	d
9.	76	34	20	52
10.	15			
11.	6195			
12.	1504			
13.	288			
14.	26.07			
15.	172900			

Page 147

	a	b	c	d	e
1.	26	90	98	87	153
2.	420	984	6062	851	15414
3.	65	50	20	14	152
4.	231	269	2478	2256	9053
5.	210	84	126	2395	4872
6.	630	14952	132348	593880	359954
7.	$.84	$5.94	$11.60	$.84	$208.80

Page 148

	a	b	c	d
8.	6	5	6	6
9.	7 r2	12 r1	51 r2	147
10.	414	907	1254	1243 r3

	a	b	c
11.	2282	184 r7	2682
12.	50	6000	
13.	9000	50000	
14.	17000	8000	
15.	6000	1200	

Page 149

	a	b		
16.	24	48	22.	3260
17.	12	48	23.	750
18.	10000	9	24.	3
19.	5	48		
20.	240	360		
21.	115	120		

Page 150

25. 16	27. 309155	29. 8537
26. 538	28. 80	30. 181 ; 7